87. 4/19/19

2.00

D0888940

THE WEATHER TOMORROW

Why Can't They Get It Right?

Bob Lynott

Published by Gadfly Press
   8925 SW Homewood St, Portland OR 97225

Library of Congress Catalog Card Number 86-83330

ISBN 0-9618077-0-9

This book was written on an Epson personal computer using
NewWord word processing program. Layout, typesetting,
and paste-up by author via modem to computer typesetting at TypeACCESS™, Portland, Oregon. Printing by
Catalogs Unlimited, Hillsboro, Oregon.

Cover photo by Dana Olsen, printed with permission of *The
Oregonian,* Portland.

"Why Can't They Get It Right" reprinted with permission of
Robin Cody.

"The Dominance of the Third-Rate Mind" reprinted with permission of American Mensa, Ltd., Brooklyn, New York, and
Alan Kent, West Midlands, England.

DEDICATION

This book is dedicated to:

John C. (Jack) Capell, who has been a pioneer in television forecasting in Portland, Oregon for 30 years (and his admirers are still counting).

J. William Wantz (1940-1985) and Charles M. Feris, who made the best weather forecast ever issued in the state of Oregon, on November 13, 1981.

ACKNOWLEDGMENT

I especially acknowledge the assorted encouragement, or assistance, or opinions of Jack Capell, Merle Pugh, Jim Westwood, Clyde Holvenstot, Carl Heeschen, Howard Graham, Joe Stein, Chet Murphy, Dean Burn, and Mike Monnie, and for stimulation from the Northwest Association of (Independent) Book Publishers, who accepted me as an associate member.

CONTENTS

1

The Unkept Promise

A generation of television babies has grown up bamboozled about the art of weather forecasting. Although viewers scoff at their favorite TV weather reporter, millions faithfully tune in the evening newscast to help make plans for tomorrow.

If the weather tomorrow is significantly different from the weather today it will upset the plans. Such changes are seldom foreseen because almost all forecasts shown on TV are made by a robot rather than a skilled human mind. If you have ever wondered "Why Can't They Get It Right?," this book will tell you.

Conditioned over the years to buy the product most often advertised, and best established in the marketplace, viewers assume official forecasts are as accurate as science can make them. After all, the Weather Service has been telling us that for 30 years. What the Weather Service really means is: official forecasts are as accurate as a robot can make them. A good forecaster can do better. This book will explain why the good forecasters are crowded out.

Proud of the Neilson and Arbitron ratings, television news directors are confident they are serving the public. We are getting what we want, or want what we are getting, or something like that. This book describes the deception, and proposes action by serious forecast users (SFUs). They are tired of being hoodwinked.

World War II demonstrated the value of aircraft. A new technology exploded in a historical sense, far exceeding the explosion of bombs on military targets.

Man truly discovered the atmosphere after he began flying in it. Weather became three dimensional. Money and talent flowed into the fledgling science of meteorology. Walt

Disney made a television movie in the 1950s called *"The Un-chained Goddess."* The mythical Meteora was glamorous.

Resources were funneled into the burgeoning science of meteorology because of the exciting promises that technology was now at hand to lick Mother Nature. Not only would man anticipate her moves, he probably would even control them, at least part of the time.

After we knew **The Weather for Tomorrow** immense economic gains would enrich us, another modern convenience would join the marvels of indoor plumbing, wall thermostats, and the TV tube. Property would be protected from loss, and **lives would be saved!**

Just give the Weather Bureau more money for research, and development of instruments, communications, computer-aided analyses, computerized calculations of future weather patterns, and of course, higher salaries. Everyone agreed the Bureau had been starved to less than subsistence because of its tradition of frugality and its lack of lobbying skill at budget time.

Since the propaganda of *"The Unchained Goddess"* the people of this nation have paid for those shiny new things. Progress has been made. Some weather services have improved. But the paramount job of **weather forecasting** is not meeting expectations. Forecasting is the most valuable product of the weather business. It depends on support services, but it is a separate activity.

As electronic technology leaped forward, the creative and intellectual part of forecasting gradually withered. Today forecasting is strangled by the monopoly and mediocrity of a Weather Bureaucracy, aided by the sedation and smugness of commercial television. The public has been shortchanged. The promise has not been kept.

What went wrong? What needs fixing?

Why This Book Was Written

A full realization of the unkept promise did not dawn on me until early 1984, several months after I protested an undeserved award to two forecasters in the Weather Service Forecast Office (WSFO) of the National Weather Service (NWS) in Portland, Oregon.

The bureaucratic reaction to the protest stimulated additional investigation, which in turn uncovered successive piles of dirty linen. I have spent approximately 6000 hours on this book. My initial concern escalated to a feeling of outrage. This is a national scandal.

I was a public forecaster (Weather Bureau) and fireweather forecaster (U. S. Forest Service) in Portland from 1949 until retirement in 1974. Why had I not understood these problems more clearly, and earlier? Partly because one within the system has difficulty seeing the forest because of the trees. Partly because exhaustion affects aging forecasters.

The next emotion was a sense of responsibility. This scandal must be exposed to public view. The task would require insight, analysis, persistence, and assumption of risk. One does not fight city hall, or challenge a federal agency on wrong doing without expectation of retaliation.

The coverup in forecasting has existed since World War II, concealed from view, not only from public view but from nearly all in the meteorological community (those in the weather business in assorted ways). The leadership of an arrogant bureaucracy, the National Weather Service (NWS), has inundated the public with misinformation and distortion.

Forecasting is supposed to predict **future** weather events. Observing and reporting **past** weather events are much easier because they are mechanical and clerical tasks. I will emphasize the distinction between **forecasting** and **observing-reporting**. The NWS strives to blur the distinction for sly reasons.

The NWS manages the non-military weather business in the U.S.A. The clerical and computerized tasks are performed satisfactorily, those which can be done by a "cookbook." But such duties are neither glamorous nor mysterious, and provide too few bureaucratic positions of power and high salary.

Therefore, with incessant propaganda and occasional dirty tricks, the NWS clings to forecasting, falsely claiming to demonstrate complex skill but actually providing a simpler routine service. If the weather tomorrow is significantly different than the weather today, the NWS will first tell you about it tomorrow. The so-called forecasts made today will be vague, ambiguous, or inaccurate.

Because I have little armor except reason (and its worth

must be judged by you), this book reproduces supportive material published elsewhere, and also excerpts from letters and tape recordings. In such material, any inserted comments by me are enclosed in brackets.

This is not a novel. It is not popular science. If you are not concerned about the weather tomorrow, lay this book down. Someday, if you are upset about a frustration in your plans because of the weather, pick this book up again.

Weather forecasting is not for the timid. Old forecasters never die. They become walking masses of scar tissue. They often are wounded before they learn to shoot accurately. Predicting the future is something like a shoot-out with John Wayne or Gary Cooper. Weather forecasting requires skill. It ought to pay well. But the bureaucrats are grabbing the money, and substituting a fake performance.

> The basic function of a teacher is to be a disturber of the peace.   (Andries Deinum, gadfly philosopher, Portland State University, retired)

2

The Philosophical Fumble

What went wrong? As a nation we turned over the weather business to the "experts." Greed and self-interest competed with science. The weather business became political. Many of the weather experts have left for rewards elsewhere. Some have seized the in-house political and financial opportunities.

Others have hunkered down to await civil service retirement, or are afraid to blow the whistle. Brave motivated experts are either retired by now, or have decided like Neville Chamberlain, circa 1938, that under existing circumstances peace is preferable to war. If any experts are unaware of existing problems, probably they no longer are experts.

Why should anyone listen to me? It will take 3 or 4 hours for you to read this book, about the time to watch a couple of movies. Then it is up to you to answer the question. Even if it is negative, I thank you for listening. I am striving to copy the techniques of writers of exposition and public discourse. Subsequent gadflies will surely raise the quality of exposition.

A gadfly is the self-definition of Socrates, a Greek philosopher executed by draft of poison in 399 B. C. for impiety and corrupting youth. A modern gadfly is a constructive critic, one who stimulates thinkers, and annoys the others.

As a young survivor of the Depression of the 1930s, I sought a "government job to gain security," becoming a Junior Observer in the Weather Bureau in 1940. Inducted into the Army in 1942 I was assigned (surprisingly, and without basic training) to an Air Corps weather station within 7 days, to help train new weather observers. The Army Air Corps eventually became the U. S. Air Force.

The Air Corps also needed aviation forecasters. World War II was fought on a global scale. Airfields were constructed in many parts of the world. This nation trained about 6000 aviation forecasters during that war.

Aviation forecasting is mostly concerned with very short range forecasts. Most flights do not take long. Accordingly, during that war, aviation forecasters were primarily concerned with analysis of existing weather conditions. Within a few hours the weather is not likely to change significantly.

I passed a course in aviation forecasting, and became an Enlisted Forecaster. I concentrated on drawing weather maps and analyzing current weather. This activity is known as synoptic meteorology, because weather maps display a summary or synopsis of existing weather. As a Staff Sergeant at the Analysis Center in Great Falls, Montana, sometimes I would go off duty and be relieved by a Major, who similarly was rated as a Forecaster. Actually, we were mainly analysts.

Although having no inside track in military bureaucracy, I never heard a shot fired in anger. After 43 months of service, some in Alaska, I was discharged as a Master Sergeant. Although I had achieved the highest grade in the 16th Weather Region (covering the air route to Russia for Lend-Lease aircraft, from Montana to Nome, Alaska) in a competitive test for rank of Warrant Officer, Junior Grade, I was passed over because I wore glasses, and hence was unfit for combat. In retrospect, it probably was because I had displayed too much independent thinking.

Caught up in the national enthusiasm for **The Weather Tomorrow**, I had no college credits for a job as forecaster upon returning to the Weather Bureau. Qualification for a forecaster's union card came in 1949, courtesy of the postwar "G.I. Bill." (It may surprise some readers to know that initially such benefit was denied to any veteran over 26 years of age, and I was 31 in early 1946.

In 1986, retired for 12 years, I have conquered most of my fear of the power structure in weather forecasting, for reasons that will become obvious.

The art of weather forecasting is so new that skill must be acquired by experience. The prescribed academic courses for a bachelor of science degree in meteorology provide useful information, but seldom offer more than 20% of the background needed for skillful local forecasting. Academic authorities place undue emphasis on mathematical skills because that is one of their off-the-shelf items. Mathematicians seldom display ability as forecasters.

Forecasting skill is limited to so few individuals at the present time, that even those in the meteorological community (which actually is small) rarely understand the process or appreciate the skill.

Because of the immaturity of this technology and its nature (predicting future events), one of the main problems is communication. Arguments and misunderstandings often arise because participants are not talking about the same thing. Their language often is ambiguous.

Hazard of the Gadfly Role

My criticism and exposure of behavior relate only to forecasting, which is only part (the most important part) of the weather business. In other environments my opponents probably would be good neighbors. Their moral and ethical convictions are as strong as mine. Unfortunately, the gulf in convictions about weather forecasting is unbridgeable.

Moral and ethical controversies should be settled in the court of public opinion, not in a court of law.

Conflict between the power structure and non-conformists began with civilization, and has existed ever since. Quoting from the *Encyclopedia Americana*, 1974 edition, condensed:

> Giordano Bruno was burned at the stake by the Inquisition on February 17, 1600. He had been imprisoned in Rome for 7 years, during which he was convicted of heresy, apostasy, blasphemy, and misconduct. He is now regarded as a martyr to the cause of intellectual freedom.

Old men are expendable. But their books, if worthy, survive in libraries.

The Scientific Method Versus the "Bureaucratic Method"

In the search for truth and excellence, the scientific method insists on open discussion and debate as knowledge is discovered, organized, and accepted. Such method is constructive. Like a navigator making a correction in course, when error is revealed an adjustment is made immediately.

Because of unopposed propaganda from the federal
bureaucracy the weather business enjoys an aura of being
scientific. The mystique of Disney's Meteora is preserved. I
regret this need to reveal that when it comes to local
forecasting by the NWS, the scientific method has been
crowded out by the "bureaucratic method." Evidence will
follow.

What Is Forecasting?

To illustrate the gulf in convictions, the ambiguity of
language, and the philosophical fragmentation within the
meteorological community, consider the definition of a
weather forecast. **A forecast is a subjective conclusion** about
impending weather.

From the *American Heritage Dictionary*:

Subjective: Existing only within the experiencer's mind and
incapable of external verification. A condition perceived only
by the patient and not by the examiner.

Conclusion: A judgment or decision reached after delibera-
tion. The outcome or result of a process. The verb "to con-
clude" means to form final judgment, to decide.

The process begins with **information** which is raw data
with little organization or analysis. Step 2 is **analysis** which
organizes and evaluates the information into a coherent pattern or picture, involving both graphic and mental images.
Step 3 is subjective and draws on memory of atmospheric
physics and past experience with similar problems.

By integrating memory, logic, and judgment the process
produces a conclusion. An accurate subjective conclusion is a
valuable product. An accurate forecast is the most valuable
product of the weather business. Measurements, reports,
charts, and messages are minor products.

Thoughtful opinions as distinguished from blind guesses
or shallow hasty judgments involve a similar process. Any of
the first two steps can borrow from the efforts of others, but
one can't borrow a subjective mental action.

Forecasting is an intellectual creative process involving
about 2% or less of the personnel in the weather business.
After such reflection the sorry state of forecasting is not
surprising.

The development of a national weather service has been absorbed in the huge task of establishing observation stations, communication and analysis systems, and administrative control over a geographically dispersed organization.

Most of the agency power structure doesn't understand the nature of forecasting. The training and background of a weather observer, or a civil service administrator, is not suited for the development of excellence in forecasting.

Civil service salaries for forecasters are less than for administrators because forecasting, as an intellectual activity, does not involve the supervision of others. Forecasters are classed as technicians. The civil service system tends to rate authority and compensation according to the number of employees supervised. I never supervised anyone during my career, and attained only grade GS-12 shortly before retirement.

A Look at Administrative Control

From the *American Heritage Dictionary*:

Bureaucracy: Administration of a government through bureaus staffed with non-elective officials. Government marked with adherence to inflexible rules of operations. Any administration in which the need to follow complex procedures impedes effective action.

Bureaucrat: Any official who insists on rigid adherence to rules, forms, and routines. The term is almost invariably derogatory.

A news item on August 30, 1984 described the noisy departure of Michael Pertschuk from the Federal Trade Commission. In a 273-page report, Pertschuk objected to the Commission's ideology and decisions. He found fault with their executive technique:

It is no accident that leaders whose ideology hardens their eyes and ears . . . reveal unlovable traits in their management style: their relationship with human beings who have the ill fortune to serve under their command or as colleagues. They tend to a heavy, authoritarian style of management: oppressive and bullying to those who work for them; servile and sycophantic to those under whose direction they serve.

They are casual with the truth. They are preoccupied with control: controlling information, controlling dissent, controlling their media images, controlling leaks, controlling emotions. Order becomes a higher value than creativity, innovation, resourcefulness, and commitment.

This reminded me to dig out the following item from old files. It appeared in *TOPICS*, Weather Bureau house organ, May 1952, page 66, author not identified. It read:

Regardless of abilities, all forecasts are issued as a Weather Bureau product and represent the combined efforts of all the staff. An individual by-line is not used, with the result that no single individual attempts to out-forecast the others. The result is a highly satisfactory, uniform, and consistently good product.

A time-worn synonym for the above attitude is "lowest common denominator." I was always trying to compete with everybody, including myself. That earned me an administrative evaluation of "unstable." The *TOPICS* item corresponded to the philosophy of the Portland boss, Eckley S. Ellison, who insisted on being addressed as the "Colonel" after his World War II service as a weather consultant in uniform, and the final rank of Lieutenant Colonel. A specific example of his philosophy is reproduced below (condensed):

Station Instruction #15, March 4, 1949, by Eckley S. Ellison

Senior Meteorologist in Charge, Portland, Oregon

Forecasts and briefings to the press and otherwise are entirely impersonal. It is not to the public interest or the interest of the station that any of these basic services should be on a personalized basis. In all cases the information should be given as emanating from the "Weather Bureau" rather than to the individual who happens at the moment to be conducting the basic services of the Bureau. The evils of having an official forecasting service identified with an individual are many:

(1) The Official-in-Charge carries the responsibility for the services rendered, and no assistant can be considered as carrying this responsibility by himself.

(2) The public learns to direct calls to the individual publicized by the press. Obviously, this is not to the public interest.

(3) The individual directing publicity to himself makes statements involving the Bureau for which the Official-in-

Charge is held responsible.

(4) Publicity is strong wine for some individuals and leads to irrational unbalanced actions. It is desired that no further offenses in this matter of personal publicity be given by any assistant in this station.

Just think! The public might learn who made frequent mistakes, or who made the best forecasts. Horror of horrors! The Colonel wanted to reserve all the publicity for himself.

Note the words "not to the public interest or the interest of the station." This revealed a shift in definition. In the bureaucratic mind the public interest is blurred with agency interest, and blurred again with self interest. (Later we shall see a 1985 memo by a supervisor at the Portland office, which also blurs the distinction between public interest and agency interest.)

Consider the following scientific attitude of Colonel Ellison, who had a GS-12 grade at the time, and was a GS-13 when he retired in 1966. In 1950 a Portland mountaineering club requested a speaker from the Weather Bureau. I was already a member of the club, The Mazamas. Ellison loaned me a copy of an earlier talk as a guide to prepare for the assignment (excerpts below):

HOW WEATHER FORECASTS ARE MADE

(to *Portland Executives*, by Eckley Ellison, 9-9-48)

It is said of Sir Isaac Newton that early in life he turned his talents to a study of meteorology, and endeavored to find the basic laws that govern the behavior of the atmosphere. He didn't do too well at it as he found the matter too complex, and soon turned his hand at simpler things such as Laws of Motion, Gravitation, Thermodynamics, and the Science of Flexions.

The rest of us have our troubles with the weather and quite often, like Isaac Newton, find the subject with a complexity that battles orderly analysis with the tools of our command. I promised that I would not dwell overlong on the technical complexities of the subject of weather forecasting, even though I may err on the side of oversimplification.

Weather is a subject that commands respect. To the airplane pilot caught in a cold front, a bad thunderstorm, or ceiling zero, he has a lot of respect for the weather. To the shipmaster in a hurricane he has a lot of respect for the weather. To the plainsman in a blizzard; to a farmer whose crop is threatened by frost; to a manufacturer on the banks of a flooded river; they have a lot of respect for the weather.

But the one who has the most respect is the one who has spent a lifetime trying to understand its vagaries. The respect, however, is of a different sort. It is rather like a profound admiration for an antagonist with the unlimited power of the universe at his disposal and who is inclined to be both the benefactor and the enemy of mankind.

My principal purpose in appearing on the rostrum is to bring you the glad tidings that the weather does make sense, that explanations can be made that really explain. The complex subject of weather forecasting can be simplified to understandable basic fundamentals.

The air mass is the heart of the new weather science. Like Einstein's Theory of Relativity, it took genius to see it, genius that itself was the product of difficult research. Anyway, we've got it roped, tied, and labeled now.

It's a great idea, perhaps the greatest in meteorology. You can recognize it as a great idea because of its beautiful simplicity. All great ideas are simple. That seems to be one of the fundamental ideas of the Universe . . .

In conclusion, I should like to ask, and answer, a question. The question: "How much difference in temperature is there between the base and the peak of a high mountain? That is, how fast does the air cool as we go to higher elevations?" In answer to this question I can introduce you to one of the rare beauties of meteorological science.

No doubt you can remember from your college physics days the simple equations of Boyle's Law and Avogadro's Law, which ordered rather complex and unintelligible data into the rhyme and reason of simple mathematics.

From these simple laws was derived the familiar and still very simple General Gas Law, the $P V = R T$ of the college meteorologist. From it is derived our simple hydrostatic equation from which the change in temperature with change in level can be computed.

We speak of this as the lapse rate. The remarkable thing about it is that it comes out exactly 1° C for each 100 meters of change. The simplicity of the relationship is the astounding thing about it.

From a knowledge of the amount and nature of the stuff that has been poured into the maw of the calculating machines to come out with the end result of 1° C per 100 meters seems uncanny.

To me, it shows that we are on the right track, for when

we do finally, with much sweat, toil, and pain, wrest one of nature's prime secrets from her bosom, it invariably turns out to be something so beautifully simple that it commands admiration and gives one a feeling of spiritual exhaltation.

It is as if the Great Architect of the Universe, when He was building, when the Earth was without form and void, He said "Let there be light", and there was light; and (I am very serious) He said "Let the temperature from the center of the Earth to the surface and beyond into the heavens change at the rate of 1° C for each 100 meters, so that when Man finds the secret he can feel encouraged."

In clumsy English measure the 1° C per 100 meters comes out to be 1.8° F per 109.36 yards. The great truth is clouded because the wrong standards of measure and the wrong approach are being used. How simple the matter becomes when the fundamental standards are used.

To me, this 1° C per 100 meters means that we do have the right standards and can proceed with encouragement that we are working steadily toward our goal in our attempt to understand nature.

To me, this 1° C per 100 meters means there is a definite order in the universe, and that we have made substantial progress toward finding the simple cosmic solution to the most perplexing and complex problem.

The reader can imagine the effect of such guidance on me, recently from the academic hot box at the University of Chicago, where I had studied for 24 months under leaders in meteorology such as Horace Byers, Walter Saucier, Herbert Riehl, and George Cressman.

Uneven Progress

The philosophical fumbling in meteorology after World War II was not limited to mundane supervisors of civil service employees. Even though imbued with the development of local weather forecasting, I also fumbled, for assorted reasons and from time to time.

In 1953, after futile efforts to obtain transfer, and realizing that simple cosmic solutions to weather forecasting would not be found in the Weather Bureau, I tried local television. It seemed an ideal medium to bring weather forecasting to the public, quickly, accurately, and conveniently.

In the 1950s the watchword in the weather world was **cooperation**. Many people didn't even know what the word meteorology meant. It needed promotion. Intramural arguments would weaken its public image. I made 325 "talks" for clubs and groups to promote weather forecasting.

However, my efforts were puny compared to the present day promotion of each local TV weather department, and presumably such promotion is nationwide. The kind of promotion I had in mind 30 years ago was to bring the art of forecasting, including frank discussion of its inherent hazards, to the public for the first time. Like a prearranged marriage, I assumed the public would like forecasting after it got to know the bride.

The promotion today is for "show business," which means to attract viewers for entertainment purposes. Program managers are willing to accept the monopoly of the National Weather Service on local forecasts, because "What have local forecasts to do with entertainment?"

Forecasts are supposed to be useful advice for managing human affairs. Maybe someday TV program managers will discover that fact. It is difficult to explain the potential human interest in genuine forecasting to someone who has only seen weather reporting on TV. Program managers don't know what you mean. They likely will ask the nearest NWS area manager.

He doesn't know either, because the ideas in this book are contrary to the administrative policies laid down by his superiors at the regional offices and Washington D. C.

In my day, KOIN-TV had no weather facilities, not even a weather teletype. Each broadcast required a long trip to the airport forecast office to study weather maps. Recognizing that weather forecasting was getting unprecedented attention via television, the Weather Bureau cooperated.

In turn, I was prudent. For TV I made my own forecasts, but when the difference from the Weather Bureau was significant, I seldom emphasized it.

My friend, Jack Capell, also resigned from the Weather Bureau, to assume forecasting duties for the Bonneville Power Administration. But his methods did not mesh with B.P.A.'s needs and procedures for long range forecasting. In December 1956 he joined KGW-TV as a weather forecaster.

The friendship and technical affinity between Capell and me, begun in the Weather Bureau, continued on television, but each tried to outdo the other. We met each afternoon at the airport weather office, and frequently had heated debates about the forecast under preparation.

Most of the time, not wanting to be outdone by the other, the one with the weaker argument would abruptly give up, and accept what seemed the better bet.

I was not suited for show business. After the 1962 Columbus Day windstorm I tried to persuade management to upgrade the weather coverage from the once-a-day 4-minute appearance five days a week. Two long letters about my proposals were ignored, even though I offered to abandon the recently acquired seasonal job as fireweather forecaster with the Forest Service.

Jack Capell was more successful at KGW-TV. He helped that station displace KOIN-TV on the ratings for news. I "coasted" on TV until early 1968 when the fireweather job was upgraded to full time.

Although disappointed with forecasting skill by the Weather Bureau, and frustrated at my failure on TV, I assumed conditions would improve when the public was better informed, and federal budgets could afford the necessities for progress. The Forest Service provided semi isolation from the Weather Bureau. Also, it appreciated my efforts.

James D. Wakefield succeeded Colonel Ellison in 1966, and George R. Miller succeeded Wakefield in 1982. I had retired, and had not been paying much attention to the rate of progress. I was loafing, and slowly writing a book on the physics of forest fire. Then in 1983, this "cause" unexpectedly appeared.

The Similarity Between Forecasting and Investigative Reporting

A reader is entitled to ask: "How come an old forecaster is making so many criticisms and recommendations? What qualifies him as an investigative reporter?"

Actually, the roles are similar. A forecaster is always striving to predict the future, something which is always hidden from view. He analyzes the present situation, and reviews the recent past. He ponders the problem and reaches a subjective conclusion, that is, a decision.

He always works in partial darkness. His analysis and logic are subject to error. He must **speculate**, which in careful usage is limited to what is considered worthy of trust, which implies his speculation is based on experience and knowledge (*American Heritage Dictionary*).

In the world of science a researcher doesn't compile data haphazardly or without hint of value. He searches where he believes truth may be found. If sufficient evidence accumulates, he may develop a theory, which must be tested and proved before final acceptance. But it is not unscientific or unethical to speculate, presume, or even to conjecture, providing the act is properly labeled. A similar technique applies to investigative reporting.

While investigating poor forecasting by the NWS, and its harassment of competitors, I wrote many letters and asked many questions. Sometimes the letters were not answered, and often questions were evaded.

A pattern of silence for a fair question about a matter of public interest suggests a hiding of something. Partial evidence, not conclusive, sometimes justifies speculation about an answer. I have been sticking out my neck ever since I became a forecaster. After the reader considers such examples presented here he is urged to make his own evaluations. He also is invited to supply additional evidence, pro or con, if available.

Another fair question from the reader: Is this exposé only a local aberration? One investigator can only cover so much ground. I live in Portland. There is much evidence that the problems in weather forecasting in Oregon and Washington are not a local aberration.

Since 1983 I have communicated with perhaps 100 individuals across the country in the National Weather Service, American Meteorological Society, and academia, and to personal friends outside the meteorological community. Criticism of NWS forecasting appears in newspapers in localities other than the states of Oregon and Washington. The ideas in this book have been tested and revised.

The problems are primarily philosophical. Good philosophy is merely common sense, an obvious misnomer, because of its lack in the world. A better synonym is wisdom, of which no one has enough.

Pros and Cons About Federal Civil Service

Because of the evils of political patronage, the establish-
ment of civil service laws and regulations brought stability
and protection of employee rights. But the system has inherent defects for tasks such as forecasting.

In spite of token awards such as step-increases within-
grade for "quality performance," and tiny cash awards for
employee suggestions, grade promotion depends more on
seniority than on professional development. The highest
salaries tend to be those in administration, rather than for
technical performance. The power structure is found among
administrators.

"Leadership ability" (which Ellison told me I lacked)
often is a euphemism for skill as a team player, which means
servility to superiors. Self-interest, a legitimate goal for ambition, is redirected to agency self-interest, which usually means
agency survival at budget time, and expansion in size, which
means number of employees and annual appropriations.

In short, civil service is inescapably political, and that is an
earmark of democracy, assuming we keep politics in public
view. It certainly is better than the raw party partisanship of
earlier days.

A civil service environment is adequate and probably
desirable for "cook book" assignments, where duties and
standards of performance are described in an Operations
Manual. Such is suitable for 95% of the tasks of the NWS.

But when it comes to local weather forecasting, which
the NWS poorly understands, such performance can best be
achieved by private enterprise, assuming that the market place
will properly reward competence. Weather forecasting will
eventually find its proper place on a par with medicine, law,
and engineering.

Continue WSFOs temporarily, but remove unfair hin-
drances to independent forecasting, including the enormous
market in public forecasting. Cease specialized forecast services to small groups with narrow specific needs. In other
words, break the existing stranglehold of the NWS on
weather forecasting.

Allow time for enticing back skilled forecasters who
transferred out (if any still exist). Allow time for developing
new talented forecasters, which was done with surprising

speed during the pressure of World War II. Then, when superior forecasting becomes obvious to the public, dismantle the WSFOs, that is, remove forecasters from the federal payroll. Early retirement benefits can be offered where appropriate.

Any real forecasters involved in the "reduction in force" can quickly find new and better careers in private enterprise.

"The Dominance of the Third-Rate Mind"

During these controversies and adversarial relationships I found an essay which provides perspective. Maybe society was unreasonable to expect a prediction for the weather tomorrow from a civil service weather agency. Maybe the nation should adjust the system, and transfer weather forecasting to private enterprise.

Here is the essay, entitled above, by Alan Kent, 17 Red Hill, Stourbridge, West Midlands, DY8, 1NA, England. It appeared in the *Mensa Bulletin*, March 1979, published by American Mensa Ltd., 2626 E 14th St, Brooklyn NY 11235-3992.

Mensa is an organization whose members have scored at or above the 98th percentile of the general population on any standard IQ test.

Alan Kent helps to understand the workings of the bureaucratic mind. Both he and Mensa have granted permission to reprint here (condensed):

> Have you ever considered the seeming paradox that well-run organizations like businesses and government departments are notable for the mediocrity of the ideas that circulate within them, when the running of organizations full of people with bright ideas, like Mensa groups, can be mediocre? It can be explained, I believe, by **The Principle of the Dominance of the Third-Rate Mind.**
>
> First, we classify minds into four groups. First-rate minds are the minds of geniuses, however these are to be defined. Second-rate minds are those of highly articulate people with considerable capacity for analytical understanding.
>
> They are problem-solvers who first formulate their problems, and later determine the limitations of their solutions, striving for better ones. They tend to be creative, and judge each creation against its purpose, rejecting it where

necessary.

People with third-rate minds are articulate but they do not sufficiently scrutinize their creations and the solutions to their problems. Indeed, their creative and problem-solving efforts can be counter-productive.

The remainder of the population, whilst varied, is inarticulate or only barely articulate, and is here lumped together as having fourth-rate minds.

Third-raters are often employed as administrators of various sorts. This does not mean that all persons so employed have third-rate minds, but the majority have.

The usual reason for a mind being third rate is an inherently limited capacity for analytical understanding. However, some persons who have sufficient understanding to be second-raters adopt third-rate attitudes, on occasions at least, because of a lack of creativity, or perhaps because of some personal, political, or religious inhibition which interferes with their judgment.

Many third-raters compensate for their limitations by an agressiveness which stands them in good stead.

Third-raters have a capacity for administrative work which makes them useful and indeed successful for a civilized society. As man evolved, the proportions of first-, second-, and third-raters must have increased from zero.

The early first- and second-raters made their prehistoric mark by technological invention. Whilst third-raters were probably useful as well at this time, they did not really come into their own until they were sufficient in number for society to "go critical" and become organized and civilized.

Second-raters are forever on the lookout for intellectual adventure, and most of them make poor administrators, finding the work boring and frustrating.

Compare a professional association, which can muster sufficient third-raters among its members for it to be well-run, with a Mensa group whose members all lie well within the second-rate range as far as mental ability is concerned, but are too individualistic for concerted action. But professional organizations are usually dull whereas Mensans rarely are.

The danger with the third-rater in administration is that while he may keep the system running smoothly by solving routine problems, he may fail to recognize, or inadequately solve, a novel major problem. Furthermore, now that civilization is established and maintained with the aid of modern techniques, there seem to be more third-raters

about than are necessary to keep it running, so that many
of them are found in situations that demand a greater ability
than they can offer.

Because the third-rater is essential to ordered society
it is difficult to avoid the consequences of his limited ability,
especially as his administrative skill enables him to
manoeuvre himself into influential positions. This leads to
the Dominance of the Third-Rate Mind. Let us see how it
happens.

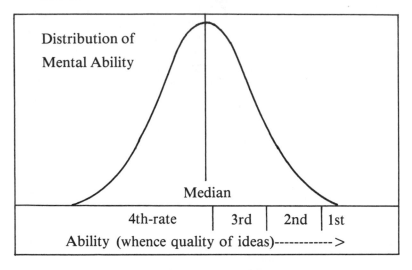

The figure is a typical distribution curve [often called
the "bell-shaped curve"] showing the distribution of some
measurable characteristic of individuals in a population
[assuming normal variation in a large sample].

[Reminder to reader: brackets enclose insertions by
Lynott]

Here the characteristic is intellectual ability and the
population is human. Ability is measured along the baseline
in the figure. The height of the curve above any point on the
baseline indicates the relative number of individuals whose
ability corresponds to that point.

The four ability groups are indicated below the
baseline, not with the object of defining them precisely in
terms of accepted measures of abililty, but to show
generally how they fit in.

Individuals formulate and support ideas [the quality of
which], statistically at least, corresponds to their ability.
[Therefore] their ability is a measure of the quality of their
ideas, especially if it concerns a fairly complex matter.

Hence the distribution curve also indicates the extent to which [quality] ideas are held by the population as a whole. [There are fewer high quality ideas because there are fewer individuals with high IQs.]

The higher the quality of an idea the greater is its credibility and effectiveness. The social impact that an idea makes depends on the number of people supporting it. Bad ideas held by a lot of people can make the same impact as good ideas held by a few. This is the essence of the third-rate dominance.

Third-raters perpetuate their dominance in any area under their control. They replace themselves by their own kind [because] they fear brighter people and fail to understand their ideas.

Third-raters dominate business and industry. Where second-raters are employed as well as third-raters this can result in intellectual inversion. The person who intellectually is the weakest member of the team, and who is therefore the most inclined to administration, becomes the leader. Thereafter all progress must lie within his comprehension. He puts a stop to innovation.

We cannot do without third-raters, but we should try to minimize the inconvenience they cause, and we should make them constantly aware of their limitations.

In our egalitarian culture these ideas of Kent may arouse emotions. However, this book is about forecasting, which rests on logic not emotion. What difference does it make right now, whether the reader or this author has a second- or a third- or a fourth-rate mind?

The point is, that forecasters must have second-rate minds. First-rate minds would be wasting their time in this technology dominated by bureaucracy. Maybe they would not in the future, after we get the brush cleared, and the swamps drained.

It is peculiar that many people glorify athletes and praise their physiques, but resent individuals with intellect.

In short, if one wants to know the weather for tomorrow, one should seek the smartest damn forecaster available. And reward her. Or him. Sometimes such performers are just as exciting as a halfback who catches a 40-yard pass.

In ancient times philosophy was presented in the form of parables, which often were misunderstood. Even with the aid of Kent's clarity, in applying philosophy to political problems one must avoid over-simplification. In breaking the monopoly of the NWS on forecasting I must describe the defects of the monopoly. However, criticism of defects does not imply total condemnation of the NWS or any individual.

Weather Service Forecast Offices (WSFOs) are killing genuine forecasting, but most other functions of the NWS are essential to any forecasting, by anybody. Also, all federal employees are not bureaucrats or third-raters.

Ambiguity About Intelligence

One must also avoid over-simplification of ideas about intelligence. Apparently, psychologists and psychiatrists consider intelligence as mostly inborn, rather than a characteristic which can be developed. A problem in such discussion is our definition of intelligence.

We usually measure intelligence by testing, and I am not objecting to that. However, I am convinced that intelligence can be increased, not just a little but a lot. It requires effort and drive, straining one's mental gut, so to speak. Mental performance also requires stamina and discipline.

Consider a more tangible human attribute, such as athletic skill. Some individuals inherit superior bodies, which give them advantage. But in our sports-minded society everyone is aware of the value of intensive training. Athletic skill does not depend only on heredity.

Intelligence, like athletic skill, is partly dependent on acquired skill. It probably is something like learning to type, or playing the piano, or pitching horseshoes.

After our culture matures beyond its present-day fascination with athletic skill, let's hope it goes on a mind-development kick. The average person has an inborn IQ potential about 50% higher than usually assumed, or usually measured. A seeming lack of mental ability may really be only mental flabbiness.

Although I have admiration for college training, academia is partly infected with bureaucracy. Higher degrees are not always evidence of mental development, or competence. In some cases they seem to be compensation for factory piece-work. (My first job in 1933 was in a rubber footwear factory, doing the printing of piece-work cards and office forms.)

One does not have to go to school to become "educated."

One does not have to be a genius to be a good forecaster, but one does have to sweat a little mental blood. It is hard on the nerves. It is exhausting. So are a lot of other difficult jobs in the world. Such as raising kids.

When it comes to forecasting, and you are in a crisis, and you are nervous and shaky, and your voice gets high-pitched, a bureaucrat might say you are unstable, and lacking in poise. Who cares, if your forecast is accurate? Such a bureaucrat probably doesn't recognize the crisis.

Conflict between the power structure and non-conformists began with civilization, and has existed ever since.

A forecast is a subjective conclusion about impending weather.

The problems (in public weather forecasting) are primarily philosophical. Good philosophy is merely common sense. A better synonym is wisdom, of which no one has enough.

The person who intellectually is the weakest member of the team, and who is therefore the most inclined to administration, becomes the leader. Thereafter all progress must lie within his comprehension.

3

The Uproar in Forecasting

Genuine weather forecasting assures immense benefits. Many benefits are presently lacking because bureaucratic profiteers have temporarily sidetracked the lofty goal of the greatest good for the largest number of people.

The technology is already here--the instruments, communication systems, and sufficient knowledge of the atmosphere. The missing link is the cadre of clever minds to fully utilize the technology. The ideal package has been described as a man-machine mix, similar to the national achievement of putting Americans on the moon. But the art of forecasting is suffering because political bureaucracy is squeezing man out of the man-machine mix.

An alarm was sounded in 1977 by Leonard W. Snellman, at that time Chief of Scientific Services in the Western Region of the National Weather Service. He said the decline in quality of forecasts was already showing up, in spite of the assistance of computers which turn out weather charts for today, tomorrow, the next day, and the day after that.

First, the computers plot and analyze existing data according to its geographical pattern. By comparing such patterns to recent earlier patterns the computer can recognize trends. Finally, using prescribed formulae, it predicts and draws patterns for the near future.

Of course, a human mind can do the same thing. After all, a human mind programmed the computer. The automated procedures are faster, and labor-saving. Also they are under centralized control. Centralized control emphasizes conformity and unanimity. It hinders local correction even when obvious error appears. Rigidity in procedure leads to the familiar downward steps toward mediocrity.

Some futurists claim the Age of the Robot is nearly here. The NWS has already assigned an utterly inappropriate job to the robots.

Just dump the weather data into a computer, press the Answer Key, and out comes a robotistic prognostic chart. Press the Key a second time, and out comes a robotistic forecast. Most forecasters with talent have left for greener pastures in other occupations.

Snellman appreciated the partly pre-cooked forecast from the computer kitchen, but he knew a gourmet dinner also required the extra culinary skill of a master chef. He wrote (slightly edited):

> Forecasters are relinquishing their input into the product going to the user. They are operating more as communicators and less as forecasters. Since this practice is increasing slowly with time, it can be called **meteorological cancer.**
>
> Many forecasters come to work, accept the computerized guidance, put this into words, and go home. Not once do they have to use knowledge and experience. This practice is taking place more and more, and it will be made easier with upcoming installation of computerized graphics and data display equipment at field offices. [now in place]
>
> Enthusiasm for forecasting and job satisfaction are declining. Current trends encourage a forecaster to follow guidance blindly, because he is criticized if he departs from guidance and is wrong, but is seldom praised if he beats the machine. [from *Bulletin of the American Meteorological Society*, October 1977, "Operational Forecasting Using Automated Guidance."]

Snellman's warning has not been heeded. Genuine forecasting is nearly dead, either from the cancer which he named, or from the chokehold of bureaucratic management, as I prefer to name it. Snellman did not explain **why** forecasters were relinquishing their input.

In my opinion, upper management does not believe local forecasters are necessary, except as figureheads to lend a local mystique of technical authority to centralized forecast "products."

Snellman was a gadfly member of National Weather Service management, not near the top in Washington D.C., but closer to the serious forecast users, the important customers of the weather business. In spite of his criticisms he was enthusiastic about "the tremendous potential for improvement in forecasting in the coming years if the **forecaster was promi-**

nent in the process.''

The forecaster is not prominent in the existing process because a civil service environment is not inclined to nurture creative and intellectual skills. Such skills are required for the art of forecasting.

If a civil service environment hinders necessary skills then why don't independent forecasters offer a competitive and superior product? Because the National Weather Service is maintaining an unauthorized monopoly. The monopoly is not absolute but it is destructive.

Not only is federal bureaucracy squeezing its forecasters out of the man-machine mix, it is squeezing independent forecasters out of public forecasting. The second squeezing is a scandal.

When voters learn the facts, a shakeup will revise the source of forecasts. Competition in the market place will quickly rejuvenate the accuracy of **The Weather Tomorrow.**

The scandal does not involve all parts of the NWS, but only the local forecasting services at WSFOs. These are the source of daily run-of-the-mill forecasts primarily disseminated by commercial radio and television.

This book is not criticizing the observations, communications, centralized analysis of data, and the generation of prognostic charts by computers. Also exempt from criticism here are warning systems of existing hazards dependent on radar or specialized observational network such as tornadoes and hurricanes. These services are essentially a matter of observation and communication.

River stage forecasts and flood warnings are not criticised because the procedures involve data management more than subjective decision-making.

Forecasting: Federal Versus Independent

The daily weather forecasts for periods up to 48 hours (with emphasis on the period 12 to 36 hours) should be produced by independent forecasters. In the jargon of the meteorological community these forecasters initially were defined as **private meteorology**, meaning private enterprise meteorology, as distinguished from **public meteorology**, meaning that conducted by the federal government.

After World War II, facing a need for unprecedented

funds for development of the fledgling technology of weather forecasting, public meteorology was seen as the victory vehicle. As a nation we had won the war; as a nation we would conquer the uncertainties of weather!

Although the uncertainty of weather obviously relates to future weather, meaning forecasting, the comparatively narrow subject of forecasting was submerged in a broader, more formal, and more impressive appellation, the science of meteorology.

Public funds could not be used to develop weather forecasting for the profit of private enterprise. But they could be spent for public weather services, and for related scientific research. The phrases: science of meteorology, scientific research, public service, national weather service, are easily blended.

Private meteorology, slighted like Cinderella, was granted the privilege to sell "specialized forecasts" to anyone not content with the "free" forecasts from the Weather Bureau. After war-time habits developed by military service, rationing, travel restrictions, and censorship of weather information, that small privilege seemed generous enough.

To this day, bureaucratic propaganda condescendingly refers to the time-honored "public-private partnership" and urges cooperation and good will between the "partners." Independent forecasters are supposed to sell forecasts in the face of totally subsidized and competing forecasts from the NWS.

Of course, there really are peculiar and "specialized" needs of certain commercial enterprises. A tiny group of independent forecasters are serving them, probably not more than 300 independents in the entire nation, including those forecasting part-time.

In spite of admission by the NWS that "specialized" forecasting should be reserved for independent forecasters, the division of turf is indistinct. For example, the government provides specialized forecasts for fruit growers in certain areas.

A little reflection suggests that many governmental enterprises and private enterprises overlap. I am not urging sharp distinction. I insist the federal government stop making forecasts for anyone, except for itself, and even then under certain circumstances only. Military forecasts seem appropriate.

The main market, the opportunity for a thriving profession of forecasting, is in forecasting for the general public, where the cost-benefit ratio is extremely favorable. Millions of people can use the same forecast.

Although the needs of some groups are peculiar or specialized, the slogan "one size fits all" is generally true. The same cold wind chills everybody.

There is no law prohibiting public forecasting by an independent forecaster, and a few are already breaching the barriers. This book is an effort to hasten the coming shakeup. The nature of the barriers will be exposed and explained.

The exposure and explanation prior to July 1985 in Portland, Oregon, were labeled a **furor** by *The Oregonian* newspaper. The defensive actions of the NWS have escalated since then. I anticipate similar resistance in other communities until the bureaucratic stranglehold is broken. The duration of the controversy will depend on the voters because the monopoly is purely a political matter.

Another example of NWS propaganda tactics is the recent ploy in referring to radio and TV weathercasters (nearly all of whom are merely reporters) as private meteorologists because they are not on the federal payroll. Nearly all of them are dependent on, and subservient to, the NWS. Very few make their own forecasts or subscribe to forecasts made by independents. Those few are subjected to pressure, both from the federal bureaucracy and from the meteorological community, which takes its cues from the American Meteorological Society (see Chapter 7).

This is an entirely new and different category from the original one for private forecasters. The federal bureaucracy always has had trouble, in varying amounts, with the private sector.

Almost everyone has heard of Irving Krick. During my career, like nearly everyone in the American Meteorological Society, I looked askance at Krick.

Now older and wiser, I can see that maybe there was more on his side of the controversies than I could see then. Aside from any question about his claims for accuracy, he challenged the arrogance of the bureaucracy.

The technique of evaluating accuracy is a large subject, too large for now.

The emergence of media weather reporters as part of private meteorology is a recent strategy to maintain the federal monopoly on forecasting. Neither the meteorological community nor the public recognizes the strategy.

This new category is as large, possibly larger, than the established category of independent forecasters. Certainly it is more visible, and being armed with the propaganda powers of radio and TV, it is a hundred times more influential with public opinion than the almost hidden group of independent forecasters.

Consider the phrase **private meteorologist**. Almost anyone working in weather can call himself a meteorologist. It is similar to the label of teacher, farmer, or newscaster. Then if one is non-federal, he or she is private, and thus the ranks of the "private sector" are expanded.

But the federal bureaucracy has also embellished the term "meteorologist" with an aura of professionalism. Media weathercasters are now recognized as **professional meteorologists**. I hasten to declare that I do not object to such designation.

After the arguments of the last 3 years, I will not criticize directly any standards that the meteorological community wishes to establish. But I will exercise my right to show those standards to anyone who will listen. Any group is entitled to call itself professional, and to establish its own standards. It is free to raise or lower those standards at will.

In the public mind a meteorologist is associated with weather, and tomorrow's weather is of prime importance. Hence a professional meteorologist is assumed to be also a weather forecaster. Well, most of them are not, and henceforth I no longer call myself a professional meteorologist, but a Weather Forecaster. My colleagues can call themselves anything they want, for example, even an Atmospheric Scientist.

I am not one of those, because I don't know enough about mathematics, and I only earned a B.S. degree. The U of Chicago credited me with 85 quarter hours of meteorology, of which 25 were for military courses. I learned forecasting during 25 years of on-the-job experience in Portland, Oregon.

Weathercasters (professional meteorologists) point to "their" forecasts, although nearly all use the "official" NWS forecasts. The new category of professional, private meteorologists in the media is a mutually beneficial arrangement.

The media reporters are professional partners of the federal bureaucracy. They "disseminate the federal products in a timely and effective manner." Both groups are assured prestige, reduced stress, and attractive incomes.

The enormous advantage to the bureaucracy is the daily access to the air-waves for the distribution of a product even more subtle and intangible than forecasts, that of self-preserving propaganda.

The reader should keep in mind that the revised definition of "private meteorology" includes both the old-time description of independent forecasters, and the new category of media weather reporters. (I concede there are a few independent forecasters on radio and TV.)

Unrest Among Serious Forecast Users

Although the NWS is happy with the status quo, and most of the public seems resigned to inaccurate forecasts, unrest is beginning to stir. Newspaper and magazine articles by writers outside the bureaucratic circle are appearing with increasing frequency. Serious forecast users (SFUs) are searching for more accurate forecasts, often vainly because of lack of local supply.

I estimate SFUs constitute about 30% of the population over 12 years of age, but the number fluctuates with current weather. That means 50,000,000 SFUs in the U.S.A., of which 1,500,000 are in Oregon and Washington. When forecasts improve, the numbers will increase. In times of bad weather nearly everyone is a serious-forecast-user.

Politicians should take note. SFUs should organize their political clout. SFUs were the ones that initiated the appropriations after World War II. They are the only real customers of the weather business. The "meteorological community" will soon be badly shaken when they learn this elementary fact.

Eliminate Weather Service Forecast Offices

Because the NWS has failed in its most important assignment, weather forecasting, WSFOs should be eliminated. Special warnings closely related to observations can be handled by the observers themselves, acting as reporters. This function is the main value of the forecast staff at present. In other words, merely eliminate the word **Forecast** from Weather Service Forecast Office.

Remove the forecast staffs and special facilities provided for them, but continue other functions. WSFOs should be redesignated as WSOs, at less cost to taxpayers. The public would then look to independent forecasters for genuine forecasting services, a product which they are not now receiving.

Newspapers, radio stations, and TV stations would be provided the same weather data from the NWS as at present, minus the so-called forecasts. Those should be provided by entrepreneurs who openly prove the value of their product.

It is strange that we look to private enterprise press services for our "news" of local affairs, but we look to a federal agency for our weather forecasts.

The reason for this inconsistency is bureaucratic propaganda that: (1) weather reports and weather forecasts are inseparable, (2) a large local staff is required for local forecasting, (3) the forecast staff must cover the 24-hour clock, and (4) no independent forecast office can afford such a budget.

The propaganda is so successful that even those working for NWS, most of them, actually believe it. Of course, a good forecaster doesn't need all that budget. A good forecast is made in a single brain, not a staff of brains.

The NWS overlooks the difference between data collection and decision-making. And if a forecast is made for tomorrow, why shouldn't the forecaster go home to bed?

During a critical period, which might need forecast revision, a forecaster might have to work overtime, like a parent with a sick child. On the other hand, when the weather is locked in a no-change situation, maybe he can go fishing.

The main clincher in the argument is that a single independent forecaster can look at the same guidance material from the same central office robot, as a whole staff of forecasters and assistants. And then he should add the missing ingredients, his local knowledge, and skill at decision-making. He will win every time there is a significant problem.

Many NWS weather observers believe that abolishing the forecast services will abolish their services. Instead, it probably will reinforce their job security. Some of the money saved from the present imitation forecasting could be used to improve their role in the national system. Weather forecasting is absolutely dependent on accurate and adequate observations, and some of these have been cut back lately. Money had to be diverted for GS-13 salaries for local forecasters to "observe" the central office robotistic forecaster.

The reader may be surprised at a claim of an uproar in forecasting. That is because it is beginning behind the scenes, as in Portland and Seattle, and it may not yet have begun in the community of the reader. The case histories described in this book should not be dismissed by SFUs outside Oregon and Washington, on the assumption that these problems are merely local problems.

Daily weather information is properly funneled through the media to the public. Newspapers usually are ready to report behind-the-scenes wrong-doing. But radio and TV stations place emphasis on "happy" entertainment rather than public education in spite of their breezy efforts to appear "informative." They are reluctant to broadcast subject matter such as in this book, especially if they feature "their own meteorologist," who almost always is under the domination of the NWS.

In turn, the radio or TV weathercaster (who rarely is a forecaster also) is busy following the wishes of the news director, and in the preparation and delivery of the weathercast. He or she is not inclined to challenge the existing system, and blaze new trails. Trailblazers seldom achieve longevity on the airwaves. Nevertheless, the change is coming soon, and maybe some of the weathercasters will read this book.

Need for Perspective

We need perspective, and one source is *"A History of the United States Weather Bureau"* by Donald R. Whitnah, 267 pages, priced at $6.00 hardcover when published in 1961 by the University of Illinois Press. The book is an extension of Whitnah's Ph.D. dissertation, circa 1942. In a review in the *Bulletin of the American Meteorological Society*, November 1962, Werner A. Baum wrote:

> This reviewer would characterize [the book] as informative rather than probing, accepting rather than critical

. . . The U. S. Weather Bureau today is the finest governmental weather service in the world, as befits the world's most advanced and wealthiest power. It has arrived at this point through the labors of many dedicated and some brilliant men. All of them were human, and there lies much of the drama.

One cannot help but feel that the story of the Bureau could be told with more excitement, with more penetrating analysis of the strengths and weaknesses of the actors in this great drama, and with more cognizance of the fact that "to err is human" for administrators and scientists as well as for Congressmen.

The paper flap on the cloth cover described Whitnah's book:

Both the fair and foul weather through which the Weather Bureau has passed are recorded in this comprehensive history of this interesting governmental agency. It is a story of political battles, power struggles, scandals, appropriation fights, personalities, and private competition, but, more significantly, it also is a story of tremendous growth, scientific and technical, and the expansion of services to the public, business and industry, and other governmental services . . .

But as ever, it still is blamed for bad weather and incorrect forecasts and rarely acclaimed for good weather and correct forecasts . . .

Due to the nature of its business, the Weather Bureau throughout its history has been the target for criticism in some form or another.

The last quotation is that of the publisher, and not that of Whitnah. The subjects of bad weather and incorrect forecasts are two separate things. The Weather Bureau was only responsible for the latter. I agree that the Weather Bureau, now renamed National Weather Service, deserves rich credit. I chose a career in that agency, and was dismayed at certain conditions which led to my resignation.

I assumed those conditions would improve as the nation developed, and in many ways they did, for example, in facilities and salaries. But in 1984, long after retirement, I became indignant and aroused at the lack of local forecasting skills, and the actions of certain bureaucrats preoccupied with self-interest.

Another source of perspective is *A Century of Weather*

Service, 1870-1970, by Patrick Hughes. In a review in the *Bulletin of the AMS,* September 1970, James A. Ruffner wrote (condensed):

> Patrick Hughes has produced [an account] of the twistings and turnings and bureaucratic reshuffling from the original Signal Service to the threshold of the National Oceanic and Atmospheric Administration (NOAA) [now in charge of NWS].
>
> Growing pains of the bureaucratic structure and change, scientific partnership and misunderstanding, forecast credibility gaps, budgetary pinches, and other foibles of any human enterprise are omitted in favor of light narrative and anecdote. The result is readable but too thin for lasting satisfaction or understanding
>
> The events cry out for interpretation, not simply statements of fact, which might be misleading. The reader is left to wonder about the personalities involved, rather than directed to a careful consideration of the difficulties involved such as the question of tested forecast skill, and other scientific and social factors of the moment, as well as the possible relevance of the psyches of the participants.
>
> What we need now is a critical analysis of the factors that produce, or hinder, a responsive weather service, in touch with social needs in peace and war, budgetary priorities, academic detachment, and atmospheric intransigence.

The atmosphere is never intransigent, which means uncompromising. Its behavior is complex, and our knowledge is incomplete, but most forecasting problems stem from lack of human effort.

In this book I don't claim to satisfy the calls for analysis by Baum and Ruffner. I merely contribute, from a viewpoint close to the boundary between forecaster and forecast user. Thirteen years after my last burden of the "weather tomorrow," my view is more concerned with attitudes and ethics than with the details of operations. I am an ex-forecaster, now a serious-forecast-user, an SFU. It will require more than one voice to get weather forecasting back on track.

The Oregonian newspaper published daily in Portland, has a Sunday circulation of about 450,000. The editor assigned a professional writer, Robin Cody, to "check out this guy Lynott and see if he is for real." Cody's article was published

Sunday, July 28, 1985, in Northwest Magazine of *The Oregonian*. Jack Hart was the editor.

Cody holds the copyright and has granted permission to reproduce the text. My cause was launched from this rocket pad.

Why Can't They Get It Right? by Robin Cody

Cody didn't have space to thoroughly answer his own question, "*Why Can't They Get It Right?*," but he interviewed individuals on both sides of the controversy. Jack Hart received unusual response via letters-to-the-editor. Some will be reproduced in Chapter 8. Rules of copyright don't apply to published statements during public discourse.

Cody began with the Pacific Northwest windstorm of November 13, 1981. During that storm, gusts reached 90 mph in Portland, Oregon, and about 115 mph on the Oregon coast.

The storm was similar to the Columbus Day storm of October 12, 1962, during which gusts reached at least 116 mph in Portland, and about 140 mph on the Oregon coast. Cody wrote (slightly condensed):

> The official forecast, issued by the NWS, called for winds of 35 to 45 mph gusting to 55 mph overnight in the Portland area. This mild non-alarming forecast went out Friday evening at 5:30 pm to television stations and newspapers, where it was passed on dutifully to the public. By that time 90 mph winds in Brookings and Coos Bay were news.
>
> The windstorm continued up the coast, took 11 lives and caused $33 million in damages. It was the Northwest's biggest blow since the Columbus Day storm of 1962.
>
> The NWS, to this day, claims to have forecast the Friday the 13th windstorm accurately and well ahead of time. The consensus among meteorologists who had no personal stake in forecasting that storm, however, is that NWS simply bombarded the major media outlets, after the storm had hit the south Oregon coast, with current-events reports. The Weather Service covered its poor forecast with a flurry of bulletins and updates that gave the impression NWS had been on top of things all along.
>
> Critics of the NWS take the Friday the 13th windstorm not as an isolated example but as the perfect illustration of what they say is wrong with weather forecasting today:

NWS is extremely cautious, slow to stick its neck out on major weather events and unnecessarily vague with day-to-day predictions. Backcasting and nowcasting eclipse forecasting.

The abbreviated forecasts in the major Northwest news media all come from the same source, the NWS. Portland television weathermen try to outdo one another with lively banter and dazzling visual aids, but they all take their cues from the same government text. The illusion, of course, is that each station generates its own independent forecasts.

This charade is what annoys true forecasters, private forecasters, who say that weather forecasts can be much more accurate and specific than what the NWS dishes out.

Says Bob Lynott, formerly the weatherman on KOIN-TV and now one of the systems most lucid and persistent critics: "Government, in collusion with the media, has lowered expectations to the point that the public is satisfied with a lot of fluff and fancy visuals about current events. Nobody expects bold forecasting anymore. People don't even realize what they are missing."

Relatively few Northwesterners know about Rich Anselmo, who lives six miles up the mountain from Oakridge, Oregon. Anselmo does the weather for KMTR-TV in Eugene, Anselmo makes his way by **emphasizing** the differences between his own forecasts and those of the NWS. He's bolder, more specific. He hates nowcasting, loves to analyze on the air why something is going to happen and to risk precise predictions.

"NWS probabilities, such as 60-percent chance of rain, are absurd--just a way to avoid responsibility and shift the gamble to the public. A fudge factor. A good forecaster has to do his homework. He has to think." says Anselmo, whose bachelor of science degree in meteorology is from Penn State.

When he's not on television himself, Anselmo, calling himself Weather-Sphere Inc., consults for public utilities and construction firms. And Anselmo, by telephone, is the meteorologist behind weather "personalities" who appear on television as far away as Memphis, Tennessee and Tampa, Florida.

"At NWS there is no incentive to go beyond computer models. NWS jobs are secure if they don't take chances. Me, though, if my track record is good, I make money. If I blow it, they switch channels or look for a better consultant."

Anselmo steps briskly from his enormous solar-heated and self-constructed log home in the woods and addresses himself to his satellite receiver dish, which is glommed onto the government's geostationary weather satellite.

"The raw data is available to everybody. The art is how you interpret the data."

Anselmo sees nothing strange in the workings of free enterprise and open competition to account for the fact that a meteorologist up the hill from Oakridge, Ore. [on west slopes of Cascade Range], has become a recognized expert on Florida freezes. Not surprisingly, Anselmo thinks Northwest forecasting, and Portland's in particular, is an embarrassment to the profession.

While Anselmo prides himself on being able to out-forecast NWS, a different attitude prevails in Portland. KOIN-TV's John Walls, without a trace of embarrassment or apology, says, "We don't forecast the weather. Even if I happen to disagree with an NWS forecast, I will go with their product."

Walls, Jim Little, David Apple and Jim Bosley, the heavy hitters for each of the weather shows on Portland's four major television channels, join chiefs George Miller and Phil Peck once every three months for dinner. "I think it's a great idea," says Walls. "We have a good, cooperative relationship here in Portland."

Walls, secure in his brightly lighted bunker two stories beneath the KOIN Tower is unmoved by the argument that free competition among local television stations might reveal true forecasting talent. Chaos would reign. "Viewers would be confused," he says.

So satisfied is Walls with NWS forecasts that he no longer bothers to keep a working file of his own on-the-air mistakes. They are NWS mistakes, not his.

"I don't have time to go back and analyze," says the man billed as Dr. John Walls, whose doctoral thesis in adult education at the University of Nebraska was titled *Visual Materials for Individual Counseling of Adults in a Community College.*

"NWS forecasts are the best we're going to get." says Walls. "They have lots of talent, a big crew, the most sophisticated data-gathering equipment."

That they do. The Portland office of the NWS at Portland International Airport hums with technological wizardry. The glow of blue computers lights the center of the workroom. A teletype prints out a fresh satellite picture every half-hour. Another teletype clacks out periodic reports and forecasts. The smell of chemicals emanates from the machine that burns off facsimile weather charts.

This office is the central processing station for weather observation posts in Oregon and offshore. Data from weather balloons, ocean-going freighters, trans-Pacific aircraft, and land-based observers are reported to the National Meteorological Center in Maryland, where a supercomputer, one of the world's most advanced, processes the raw data.

In return, national headquarters sends the Portland office a broad, general, computer-generated prediction of weather to come. Six "lead" forecasters analyze the big picture, adapt it to local conditions and make weather forecasts for 12 regions in Oregon.

George Miller, chief meteorologist at the Portland station, defends NWS as a collection of the best scientific talent available. The problem comes in converting the big computer models into detailed local forecasts.

In computer models, weather happens in a smooth, cooperative, average sort of way. But on the ground, weather arrives — like dandelions — in sudden local outbursts.

Miller concedes that a private forecaster can tailor information more accurately than NWS to a particular client. "We're not highly specialized here." But he is skeptical of NWS critics who focus on the adverse more than on the positive.

"What's happened," says Miller, "is that forecasting skills shot upward with the advent of computers in the 1950s and satellites in the 1960s. Now we have reached a kind of plateau where forecasting has not improved as fast as many people expected."

The graph line soon will be propelled upward again, Miller predicts, by new supercomputer models that will read atmospheric measurements at 16 vertical levels, instead of the current six. If the mesh is tighter, fewer surprises will slip through.

In response to criticism that his staff has been too conservative when it comes to venturing its own forecasts, Miller will not be dragged into anything more than repeating "This staff is professional."

"In terms of fewer misses," says Miller, "studies show that the general trend is down."

Fewer misses.

"Fewer chances," says Bob Lynott. "NWS forecasters will consistently opt for the security of not being wrong over the thrill of being right.''

"That's the key,'' agrees Charlie Feris, the meteorologist for the Bonneville Power Administration. "At NWS, weather is just a job. There is no incentive for taking risks.''

Lynott talks about the snowstorm that tied up Portland ground transportation last December 20. News reports the next day said forecasters were caught by surprise.

"Of course they were all surprised,'' says Lynott, "because all the TV guys were reading off the same wire from NWS, which called for 40% chance of rain or freezing rain.''

Lynott pauses to see if his listener can figure out how cautious that really is. "Forty percent,'' he says. "Rain OR freezing rain. How vague can you get? No mention of snow,'' he says.

"They did put out a Winter Storm Watch, not a warning, which means NWS was prepared to **watch** that storm along with the rest of us and tell us what happens.

"That they did,'' he continues. "By the time NWS revised its forecast to include the possibility of snow for Portland, it was already snowing in Scappoose.'' [15 miles northwest from Portland]

Nowcasting.

"It's like rifle marksmanship,'' says Lynott. "The target gets closer as time passes. If you wait to pull the trigger, eventually the target touches the muzzle, and a bull's eye is easy.''

"Lynott's right,'' says Pat Timm, president of the Clark County Amateur Meteorological Society. "They're real careful at NWS.''

Timm, 35 and very mild, mostly sunny, was forecasting weather for Cox Cable TV in Vancouver as the Dec. 20 snow approached. Videotapes show Timm, the amateur who forecasts weather as a hobby, predicting snow two days before it arrived.

Timm is the first to admit that he is wrong sometimes. And in most cases he does agree with the NWS short-term forecasts. "But it's pretty hard to believe they missed that one,'' he says. "It was snow all the way.''

"Forecasting,'' says Timm, "is not a science. A forecaster needs the scientific knowledge, but forecasting itself is an art. Some people have a knack for it, like playing the piano. Then you have to be willing to stick your neck out.''

Is **anybody** forecasting on Portland-area TV?

"No,'' says Timm, after a long pause. "Maybe Jack

Capell on KGW-TV, weekends. But no, it's a shame."

William P. Lowry is a Certified Consulting Meteorologist who lives in McMinnville. Lowry, 57, got his master of science degree in meteorology from the University of Wisconsin and his doctorate at Oregon State University.

"First," says Lowry, "the NWS has made great contributions to meteorology. NWS has good scientists, the funding for research, and the machinery — satellites and computers — to study the atmosphere. Nobody argues with that.

"But there is an important fuzzy line between the technician and the forecaster," says Lowry, who theorizes that forecasting skills have languished because of technological advance.

"The technological fix is in," he says. "Very little forecasting comes from the guy on TV, or the guy at the airport. What little we get comes from the National Meteorological Center near Washington D.C. This product is called guidance. With no incentive to interpret it, guidance becomes the forecast."

Fred W. Decker, a former U.S. Weather Bureau forecaster who taught physics and meteorology at OSU from 1946 to 1981, writes: "The NWS needs reorganizing and refocussing to do what it can and should do best."

The responsibilities of a reorganized NWS would include, Decker says, all the common functions leading up to actual forecasts. But, he concludes: "The federal weather service would do well to withdraw from general weather forecasting."

McMinnville's Lowry says: "Meteorologists like Bob Lynott and others are worried about professional integrity. Lynott has kept his tools sharp. That's why he is so concerned about high rewards going to those who have not."

Bob Lynott. Remember Bob Lynott? When Robert E. Lynott left the U.S. Weather Bureau's Portland office and talked himself into a part-time job at KOIN-TV in 1953, he became the first television weather forecaster in Oregon, one of the first in the world.

A whole generation of Northwest viewers know how Lynott could raise his black grease pencil, and with one deft sweep, splay a perfect dotted line across a Plexiglass map. With his Mr. Peepers glasses and his undisguised intellectual vigor, Lynott was charming when he was right and absolutely infuriating when he was wrong.

Often mistaken, but never in doubt, Lynott, with his

bachelor of science degree in meteorology from the University of Chicago, considered himself a teacher.

And if farmers, loggers, and vacationers were more likely to focus on his mistakes than to share Lynott's relish for the learning process, his enthusiasm was always there. Lynott projected an unmistakable sense that if he didn't predict the weather right yesterday, by gawd he would get it today.

Although no one knew it at the time, the 1950's were something of a golden age in meteorology. World War II aviation had given new urgency to the study of atmospheric conditions. Government and the post-war aviation industry attracted fine minds to meteorology.

There arose new hope that man's scientific method would be of wonderful new utility in reporting the weather and even — to the most euphoric thinkers — in changing the weather. Research monies flowed. Expectations were high.

"Meteorology has come a long way," says Lynott, "but forecasting has not improved in 20 years."

Jack Capell, who became KGW-TV's weatherman in 1956, shares Lynott's disappointment, if not his anger. "No, we weren't better in those days," says Capell. "Our tools were very primitive. But I am not impressed with any recent improvement."

Capell, like Lynott, is a former U.S. Weather Bureau employee. Now 62, he is crippled and confined to a wheel chair. Capell speaks to the accompaniment of his soft, wonderfully radiant blue eyes, speaks carefully and thoughtfully, trying to be fair.

"Lynott and I could explain it any way we wanted," says Capell. "The difference now is that satellite pictures and good graphics, while they enliven the telecast, are not forecasts, and they take time. Giving more and more current statistics takes time. Bob and I used that time to make the public familiar enough with the principles that they could think along with us.

"Admittedly," says Capell, "we addressed ourselves to Portland-Vancouver and let other areas, Eastern Oregon, the coast, infer from our description what they could expect. Unless something important or different was due there.

"The other big difference," he says, "is that Bob and I were out front taking the heat whenever we were wrong. We were wrong many times, of course. But we couldn't hide, and we got better. We were in competition with each other. Today, NWS forecasters are anonymous. Anonymity tends toward mediocrity."

After Lynott left KOIN-TV in 1968, and retired from his job predicting forest fire weather conditions for the U.S. Forest Service in 1974, he settled in the high chaparral country east of the Cascades near Culver, Oregon. Lynott planned to take up gardening and fishing.

The trouble was, his mind kept working. And that's a wonderful and terrible thing to happen to a person, because when he finally has the time and the distance to see the whole pattern, he has lost the position of influence he needs to do anything about it.

One day in 1983 Lynott reached into his rural mailbox and plucked out a fresh *Bulletin* from his beloved American Meteorological Society. Inside, he read that two Portland NWS forecasters had been nominated by Portland NWS, and had won, the prestigious Charles D. Mitchell Award for exceptional forecasting of the 1981 "Friday the 13th Windstorm."

Lynott was astonished. So were others.

"My teeth nearly fell out when I read that," said Feris, the BPA meteorologist. Feris and Bill Wantz, at BPA, are generally credited with having had a much earlier and more accurate handle on that storm than did NWS.

BPA put out preliminary warnings to its dispatchers at 9:30 a.m. Friday the 13th, with final warnings at 1 p.m., and accurately predicted higher winds than NWS ever did.

"Feris and Wantz were on it," says Capell. "NWS didn't do a very good job."

For Lynott, this miscarriage of official recognition, the Charles D. Mitchell Award to NWS forecasters, was the catalyst that brought him storming out of retirement. His cause: "To break the stranglehold of the National Weather Service and to revive the art of forecasting."

From his outpost near Culver, Lynott has mounted an all-out campaign to arouse the public to the notion that if forecasters were "privatized," the public would be far better served.

Lightning bolts issue from his hot typewriter: "Television has bought into a huge, lurching, bureaucratic monopoly that refuses to take risks and has become expert at covering its own ineptitude." Lynott calls NWS a "two-bit agency, an appalling intellectual wasteland."

A shrillness of style sometimes gets in his way. Although he has had a few short thunder rolls printed in newspaper letters-to-the-editor pages, Lynott's passion and combative nature make it fairly easy for his critics to dismiss him as a curmudgeonly old crank.

Another problem is that Lynott's thesis is outrageous to a public already anesthetized into thinking the way television would have it think, briefly and fuzzily, about weather. Lynott's listeners see his blue clouds of rage and dismiss the possibility that he might be a clear thinker and a reputable meteorologist, which, in fact, he is.

Lynott, undiscouraged, has established his own Gadfly Press and is hard at work on a book, "The Weather Tomorrow."

Among Lynott's supporters is Chuck Wiese (pronounced Wise), now the weekend weatherman for KING-TV in Seattle. Wiese, 28, was the precocious weather-kid on Portland's KPBS radio at age 12, moved to KOIN-TV at 15 and later received a bachelor's degree in atmospheric science.

When he came back from college to KOIN-TV, he ran headlong into John Walls' policy of not varying significantly from NWS forecasts.

Because he measures his own professional success in terms of his ability to "scoop" NWS forecasts, Wiese is something of a lightning rod to those who argue that forecasting belongs in the private sector, as well as those who disagree.

"He (Wiese) was terrible." says KOIN-TV's Phil Volker. "He was wrong all the time. Always alarming people."

"Chuck Wiese is a top-notch forecaster," says BPA's Feris.

At KING-TV in Seattle, naturally enough, Surges Dorrance, general manager, is "very excited" about the station's effort to develop its own independent forecasts. No other Seattle station is doing so. The KING-TV team of Jeff Renner, Chuck Wiese, and Larry Schick frequently differs from NWS, according to Don Varyu, news director.

Verification studies turn out to be mostly cloudy.

Clifford F. Mass, assistant professor at the University of Washington, unleashed his Atmospheric Science 101 students in the spring of 1984. They watched every weather program in Seattle and compared individual forecasts to one another and to the NWS.

Wiese says the study showed KING-TV weather was the best. But Mass says, "There was no statistical significance to who was best. You'd need more time, a larger sample."

Professor Mass knows of no scholarly work that has

been more successful than his own attempt to quantify and to score forecasts. He adds: "I was afraid the stations would use my results to claim they were better than NWS. Private sector vs. public sector. A whole bucket of worms."

Which prompts Wiese and others to mutter darkly about where Mass' funding comes from — the public sector. But the lesson here is that forecasting, like the weather itself, is not easily boxed and scored. The potential for "proving" what a forecaster wants to believe about his own performance is enormous.

George Miller, Portland's NWS chief, can say, "We did forecast snow for Portland (on the day before) for December 20."

From deep in the files he pulls out a December 19 special statement about the "possibility of freezing rain or locally heavy snow . . . for the Columbia River Gorge, Portland urban area and Northern Oregon Cascades." There it Is, all covered.

Yet NWS forecasts later that afternoon and evening, specifically for the Portland area, continued to call for rain or freezing rain, 40-percent chance of either.

A generous interpretation would be that NWS predicted everything but oobleck for Portland, and therefore was not wrong.

"What matters," says Lynott, "is the changes, the big weather, and that's where NWS backs off."

"What matters," says Wiese, "is that there are people around who have proved their ability to recognize big weather before it gets here, and then to step out front and nail it. You think Portland couldn't use that?"

Of particular concern to Lynott and Wiese — and to Anselmo in Oakridge, Timm in Vancouver, Lowry in McMinnville, Feris at BPA — is that the government's nothing-ventured-nothing-lost forecasts have become standard television and newspaper fare. The public thinks that is what a forecast is.

The difference between these meteorologists and the bewildered viewer who watches a TV weather show and wonders why his head went numb — why he can't remember the forecast — is that the experts know why.

And they know it doesn't have to be that way.

Robin Cody and I were strangers until our first meeting March 15, 1985. Cody demonstrated unusual ability to quickly grasp the complex problems which *The Oregonian* was suspecting, and of which the television industry both locally and nationally seems unaware.

Upper management in the NWS does not believe local forecasters are necessary, except as figureheads to lend a local mystique of technical authority to centralized forecast "products."

Serious forecast users should organize their political clout. They are the only real customers of the weather business. The "meteorological community" will soon be badly shaken when they learn this elementary fact.

It is strange that we look to private enterprise press services for our "news" of local affairs, but we look to a federal agency for our weather forecasts.

A good forecast is made in a single brain, not a staff of brains.

4

Phony Award for Exceptional Forecast Service

Taxpayers must be persuaded that the art of forecasting really needs reviving, and that the NWS is an obstacle.

The annual budget for the Portland WSFO is nearly $2,000,000. The basic salaries paid to Portland "forecasters" are estimated near $650,000. The duty of forecasters is to predict future weather, especially weather which threatens **life and property**.

This chapter describes an event which the NWS regarded as the finest example of weather forecasting performed by the NWS in the entire nation during the year 1981. The American Meteorological Society agreed, and bestowed the prestigious Charles D. Mitchell Award on two Portland forecasters for "exceptional forecasting service that was accurate, timely, and effective." (*Bulletin of the AMS*, January 1983, page 626)

We can use their yardstick, and examine their standards of professional performance on a day when they claimed their skill was remarkable. As you will see, this is a classic example of the ambiguity of language in the meteorological community. I will be talking about forecasting. NWS personnel will be talking about reporting and communicating.

If a forecast is accurate, and timely (providing sufficient advance notice), it is someone else's responsibility to make it effective. The repetitious use of the word **effective** reveals something about the mind-set of the federal bureaucrats in the NWS. This word must be in a propaganda manual.

The windstorm of November 13, 1981, in the Pacific Northwest was the area's third worst windstorm in recorded history. The No. 1 storm was on January 9, 1880, but hardly anybody knows about that, and comparatively few people were harmed because the population was sparse.

The so-called No. 1 windstorm was on October 12, 1962, nearly 83 years later. Gusts reached 116 mph in Portland. All forecasters, including this one, were overwhelmed by an event beyond their understanding.

Owen P. Cramer, research meteorologist for the U. S. Forest Service, and I made a post-mortem of that storm, which was published by the Weather Bureau in *The Monthly Weather Review* for February 1966. The mystery was solved. Although awesome, the characteristics of such storms should now be clearly understood.

About three storms of comparable intensity occur in the Gulf of Alaska each year, on the average, and follow a left-curving path. Only occasionally one swings close enough to the land areas of the Pacific Northwest to seriously endanger life and property.

Forecasting is a risky endeavor. A serious-forecast-user (SFU) must trust the judgment of the forecaster. Is the NWS worthy of that trust? Is it worthy of the budget for local forecasting?

Report on the Pacific Northwest Windstorm, November 13, 1981

An official report with the title above was issued by the Western Region Headquarters of NWS in Salt Lake City in May 1982. The report is a whitewash so shameless that presumably it was intended for internal agency use only, to help administrators cover up forecasting incompetence from congressional view.

If it is not a coverup, then the NWS cannot distinguish between forecasting (ahead of time) and reporting (after the time). Copies of the report, possibly no longer available from NWS, can be supplied at cost plus handling (Gadfly Press, 14 page paste-up, 8x14 $5.00 postpaid).

Considering the advances between 1962 and 1981, such as state of knowledge, increases in staff and facilities, and improvement in communications and guidance material, the NWS performance in 1981 was a greater fiasco than in 1962. Tax dollars for local forecasting are being wasted.

Damage during windstorms is related to peak gusts which endure briefly, say from 10 to 20 seconds. Any shorter are almost impossible to measure. The NWS euphemism of "sustained winds" refers to the **average wind speed** over a period of at least one minute.

Peak gusts are approximately 50% faster than one-minute average winds. They occur when faster winds from aloft, winds not yet slowed by friction on the ground, dip briefly amid the turbulence. Gusts may be compared to an airplane swooping down for a landing.

In 1981 peak gusts were about 90 mph in the Portland area (where nearly one million people live), 110 mph on the Oregon Coast, 125 mph on small mountains inland in western Oregon, and 80 mph in north central Oregon.

Winds were less strong in Washington because the storm was slowly dissipating as it moved northward just off the coast. Gusts were measured at 67 mph at the Seattle-Tacoma airport.

Garbage-Can Forecasts

The final forecasts before the winds arrived, issued by the Portland WSFO about 6 p.m., predicted winds of 35 to 45 with gusts to 55 mph in the Portland area, and 40 to 60 knots with stronger gusts (how much stronger?) for the Oregon Coast.

I was unable to obtain the complete forecasts issued by the Seattle WSFO, but at 9:30 p.m. the local forecast for Seattle was for winds 25 to 40 mph with higher gusts (how much higher?), and for 40 to 50 mph with higher gusts (how much higher?) for the coastal strip of Washington. Both forecasts were introduced as high wind warnings.

If taken literally, such forecasts would inspire only mild concern at that time of year in the Pacific Northwest (move garbage can inside garage).

Unsung Heroes

William Wantz and Charles Feris, federal forecasters for Bonneville Power Administration in Vancouver, Washington (semi-isolated from NWS influence) had expressed deep concern to NWS as early as 9 a.m. on November 13. About 1 p.m., unwilling to wait any longer for a "coordinated" forecast, they had issued a wind warning to electric utility companies for winds gusting to 90 mph on the coast and 70 mph in interior valleys.

Wantz and Feris deserve praise for the best weather forecast ever made in the Pacific Northwest. But they were not honored by the American Meteorological Society, although they were nominated belatedly by Chuck Wiese. Presumably, that would have weakened the earlier "honor" bestowed on the NWS.

William Wantz died of a brain tumor November 10, 1985. I had promised him this book would recognize that forecast.

Excerpts and Comments

Excerpts from the NWS **Report** are shown here in miniprint, with short comments by me enclosed in brackets. Excerpts may be abbreviated, but meaning will not be changed.

It is customary after a significant weather-related disaster for the NWS to evaluate the effectiveness of the warning system. Fifty million dollars worth of damage and eleven deaths were attributed to this storm.

In accordance with NOAA [National Oceanic and Atmospheric Administration] Directives Manual 2 8-17 and NWS Operations Manual, Chapter 1-06, a Survey Team was dispatched to determine performance of NWS offices during the storm. The team consisted of:

Richard Hutcheon	Asst. Chief Met. Services	Regional Hdqtrs
Art Hull	Meteorol. in Charge, WSFO	Seattle WA
Jim Wakefield	Meteorol. in Charge, WSFO	Portland OR
Dale Goudeau	Acting M I C, WSFO	San Francisco CA
Charlie Hanas	Official in Charge, WSO	Olympia WA
Seymour Krepky	Official in Charge, WSO	Astoria OR
Vernon Mattox	Official in Charge, WSO	Salem OR
Nile Woltman	Meteorol. in Charge, WSO	Eugene OR
Lyle Hammer	Meteorol. in Charge, WSO	Eureka CA

This was equivalent to sending foxes out to investigate disappearance of chickens. To determine responsibility for language in report, Chuck Wiese asked Richard Hutcheon if he wrote the Report. Hutcheon claimed all of Team wrote report. Richard Hutcheon is now Area Manager at WSFO Seattle, Washington.

It became apparent, immediately after the storm, that the NWS had performed in an excellent manner. It presented a unique opportunity to determine the value of weather warnings when the NWS's warning system works in a most effective manner.

The Team interviewed state, county, and local officials, special users, and the general public to determine timeliness of the warnings, response taken, and the resulting savings.

Interests, where property was vulnerable and preventive measures could be taken, acted to protect their property before the high winds began. Other groups, which have responsibility to respond to problems caused by high winds [such as emergency crews] took actions to be ready to respond in the most effective manner possible.

In addition, some of the people interviewed stated that even though they were unable to take action, knowing what was going to occur gave them a certain "peace of mind."

This is an example of propaganda intended to plant ideas by assertion. **"Knowing what was going to occur . . ."** How could people in the Portland area be given peace of mind about impending gusts to 90 mph with only a high wind warning of 35 to 45 with gusts to 55? They secured their garbage cans, with the trustful peace of delusion.

The report reveals the agency's concept of its "warning system" for high winds, which supposedly worked in a "most effective manner." As a forecaster, I would assume a warning in this framework means advance notice of the wind, with an accurate estimation of peak force. A high wind warning should be a forecast, not a weather report after the fact.

But forecasting is much more difficult, and more risky, than reporting. The NWS is skilled at observing and reporting. To obscure its lack of skill at forecasting, the NWS strives to blur the difference between reporting and forecasting.

Forecast accuracy refers to the correlation between predicted events and actual events. It requires advance notice because the value of a prediction decreases to zero as the advance notice decreases to zero.

The NW uses the word **timeliness** in an ambiguous manner. It declares a forecast is timely, and implies the forecast gave sufficient advance notice of the event. According to the dictionary, the word timely may have some relation to advance notice, but the primary meaning of the word is "occurring at a suitable or opportune time."

Opportune means suited for a particular purpose, such as an opportune meeting. It means occurring at a time that is fitting or advantageous. A timely warning may merely inform you of existing hazards, such as the familiar travelers' warnings for snow already falling at mountain passes in the Northwest, snow not yet visible from your home in the valley.

Ambiguity is a refined art among WSFO forecasters and managers. The NWS has long experience with dissembling, and in collecting favorable testimonials. It seeks "cooperative feedback." The NWS subtly informs its "users" that criticism of services, or non-cooperative feedback, will in turn reduce NWS "cooperation" in providing those services. And where else can users obtain weather services, under NWS monopoly?

> Improvement in NWS watch/warning programs are dependent upon cooperation and feedback from public officials, news media, special users, and the general public.
>
> The NWS performed at a very high level of effectiveness during this disaster, and this performance reflected favorably upon NOAA, the NWS, the Western Region, and the stations in the Pacific Northwest.
>
> Comments received by the Survey Team after the storm from people in the affected area indicated a high level of satisfaction with regard to the warnings. Many indicated that the NWS could not have performed better and that the wind forecasts were the most accurate and timely in memory.

Several factors combine to help the NWS collect the answers it wants. When the storm is raging, the NWS is disseminating warnings intermingled with reports of violent weather. The agency appears to be on top of things. The average citizen is unaccustomed to analyzing time sequences.

For many days John Q. Public remains distraught by the event, and is flattered by the hype and salemanship of an official survey team. The individual is unlikely to be critical of a "hard-working" agency coping with an act of God, especially when he doesn't understand what good forecasting should be.

The Report began with an Executive Summary, which included 1½ pages of Findings and Recommendations. These pleased the author so much he repeated them verbatim at the conclusion of the Report.

The word **finding**, used as a noun, is a new bureacratic buzzword. It seeks to establish an aura of judicial authority to a self-serving assertion.

In *The Oregonian* of January 19, 1987, Mary McGrory, of Universal Press Syndicate, ridiculed the word **finding**. She said, "Writing a *finding* is a tempting way of life. Write yourself a memo, saying you're doing it for a higher purpose, and then go out and rob a bank."

David Broder, national columnist, in the same paper two days later, volunteered to serve as acting president of a non-profit group called Safeguard America's Vital Euphemisms (SAVE). He claimed facetiously that without euphemism, neither America nor its government can survive.

Six of eight findings in the storm Report are listed here. Number 6 is brief and acceptable.

Finding #1. The NWS performed at a very high level of ef-fectiveness. In general the warning system worked. Recommendation: Appropriate recognition should be given to NWS offices and individuals whose performances were well above the norm.

Finding #2. The offshore anchored buoys provided in-valuable information regarding the track and intensity of these storms, and combined with satellite imagery, provid-ed the information forecasters needed to issue warnings well in advance. In particular, the buoys provided critical, timely information that was not apparent through National Weather Center analyses or forecast guidance.

The NWS is entitled to establish its own standards. It claims the forecasts from the Portland WSFO were above the norm. Thus the norm is even lower.

Why did the invaluable information from the buoys make more of an impression on Wantz and Feris than it did on the two NWS forecasters, Ron Surface and Jerry Youngberg? Probably Ron and Jerry were squeezed out of the man-machine mix.

Finding #3. The satellite loop obtained from a Seattle TV station played an important part in the decision by forecasters that the storm would curve northeastward and bring winds over western Washington . . . It was brought to WSFO by Harry Wappler from KIRO-TV about 2:30 p.m. and was studied in detail by the oncoming swing-shift forecasters. The track of the storm showed a much more northerly component than forecast guidance indicated.

Harry Wappler is a television reporter who faithfully

and articulately promotes NWS "products," and in turn, receives NWS promotion that he is a professional meteorologist, which he is, according to NWS standards. But he is not a weather forecaster, according to my standards. He is a good weathercaster, and I hope someday his station will provide him with non-federal forecasts.

The track of the storm as shown on the loop film was after the fact. The upper air patterns at 8 a.m. clearly showed steering toward northwest tip of Washington, which was the basis for concern of Wantz and Feris. It also worried Ron Surface, but apparently he was obliged to write his forecasts according to guidance from the robot, which at that time said the storm would move inland to the south of Portland.

> Finding #5. Power outage, particularly at WSFO Portland, reduced the capability of the NWS to keep the public informed [about current events]. While most forecasts and warnings were issued in a timely manner, even during the power outage [thus "timely" means right now], the method utilized was awkward and time consuming, requiring extra assistance from Seattle and Medford when they were also involved in a warning situation. [when the wind begins to blow dangerously]

> Finding #6. Most National Weather Radios in the Pacific Northwest were off-the-air because of power outages. Recommendation: Ensure that all transmitters have emergency power. As emergency officials come to rely on National Weather Radio as their primary means of obtaining weather information, the need to keep NWRs operating in severe weather becomes increasingly important.

> Finding #8. National Meteorological Center guidance greatly underplayed the intensity of the storm. The performance of the numerical guidance points out the need for forecasters to integrate the numerical guidance with other information such as careful interpretation of observations. These were the keys to catching the development of this storm. Recommendations: The concept of including and even increasing the forecaster's role in the decision-making process with regard to issuance of warning must continue to be promoted.

For this storm, the NWS "had lost its keys" and bureaucratic management "didn't know where to find them."

How can a concept continue to be promoted, when

always it has been demoted?

Finding #8 suggests the underlying embarrassment of the Survey Team. It echoed Leonard Snellman's appeal for a forecaster's contribution to the man-machine mix, and pretended the concept was alive and well.

The computerized forecasting system is not programmed for rapid significant changes. Such changes are the reason for the need for human forecasters. If significant changes are not impending, then anyone can rely on persistence. Tomorrow's weather will be similar to today's weather

The double-talk about forecasts and warnings was transparent. Consider these statements:

> Without question, the timeliness and quality of the warnings for this storm were outstanding.

> No problems occurred with regard to warning dissemination prior to the onset of the strong winds. However, once the winds hit, the massive power outages had a significant impact on some offices ability to operate effectively. Power at WSFO Portland went out at 3:24 a.m. This left Portland without communications except by telephone. Personnel remained at the Portland office, prepared messages and dictated them to Seattle and Medford over the telephone. These offices, in turn, transmitted them on various communication circuits. During the power outage, only a few forecast products were missed . . . During the period of the Portland power outage, WSFO Seattle had extra personnel on duty to coordinate with WSFO Portland personnel and to assist in dissemination of forecasts and warnings.

While the winds were raging, the NWS was valiantly striving to distribute its **reports,** to tell the public the wind was blowing. This reveals the mind-set of the NWS, to observe and to report. The power outage interfered with its job of reporting.

How can a **forecast** for high wind be issued when the wind is already blowing? The power outages occurred after the target touched the muzzle of the forecasting gun.

These statements reveal that the NWS defines a **warning** as a notice of dangerous weather already occurring. Also, if the warning is prompt, the NWS claims it is "timely." Promptness in reporting is not prediction.

In summary, the **1981 storm** was exceptional. Starting

with self-delusion within the NWS, and with clever dissembling to the public, many were convinced the **forecasting** was exceptional. That included the leaders of a respected scientific organization called the American Meteorological Society. How did that happen?

The Abuse of the AMS Award System by the NWS

This shabby story begins at the Portland, Oregon, WSFO on March 1, 1982 while the Western Region Office of the NWS was preparing the whitewashed Report on the Pacific Northwest Windstorm. James Wakefield had already retired, and George Miller had not yet arrived to become the new Meteorologist in Charge, and Area Manager Philip A. Peck, the supervising forecaster, was the Acting Area Manager.

The chain of events reveals the stranglehold on forecasting by the NWS is not merely a local aberration in Portland, Oregon.

Peck prepared a 2-page nomination of Ronald K. Surface and Jerry A. Youngberg, "lead" forecasters at Portland, for the Charles D. Mitchell Award given each year by the American Meteorological Society.

The award is for individuals who, through performance of exceptional forecast service, have distinguished themselves by an exceptional specific prediction, and brought credit to their profession.

The nomination was sent to Harry S. Hassel, Deputy Director at the Regional Office in Salt Lake City. The Director was Hazen H. Bedke, now serving a three-year term for a second time on the Council of the national AMS. He had served as Chairman of the national Committee on Professional Ethics and Standards in 1957-58.

The nomination was endorsed, presumably forwarded and endorsed by NWS administrators in Washington, D.C., and submitted to Dr. Dale A. Lowry, Chairman of the Weather Forecasting and Analysis Committee of the AMS. (Dr. Lowry worked in the Techniques Development Laboratory of the NWS in Silver Springs, Maryland.)

Lowry's committee recommended the Award, which was presented to Surface and Youngberg at the AMS Awards Banquet in New Orleans, Mississippi, on January 12, 1983. It appears that NWS self-interest dominated the procedure.

This was part of the whitewash of NWS forecasting for the 1981 storm. It reveals the bureaucratic influence of the NWS on the AMS. This influence will be difficult to purge.

After learning in July 1983 about the Award, I wrote to George R. Miller, new boss at Portland WSFO, requesting copies of forecasts made on November 13, 1981. They arrived after 28 days. The request seemed to generate a defensive attitude by Miller, an attitude of "circle the wagons, men, we spy an Indian on yonder hill " Miller declined to identify the nominator for the award, which I also had requested. He wrote:

> Bob, I must confess, I do not like the tone of your letter. Rather than a request for data, it appears to be an attack on one of the forecasters at Portland.
>
> I hope I am wrong, for I consider an attack on anyone who works in this office as an attack on myself. The recommendation was made prior to my arrival. However, I agree with it totally.
>
> Everyone I have talked with since coming to Portland has had praise for the NWS regarding this particular storm. Radio stations, television stations, newspapers, Coast Guard, harbormasters, etc., have spoken very favorably of our actions.

I was investigating the **forecasts** made before the storm, not the **actions** of the NWS during the storm.

Three days after the storm *The Oregonian* printed an editorial "No-Name Storm Gives Less Warning." It read:

> A major part of the problem is that the Pacific Northwest has a nondescript [a gentle adjective] early warning system for storms. Considerable damage and possibly some lives might have been saved if Oregonians had received a clearer mental image Friday night of the kind of blow coming their way.

I prepared a detailed formal protest of the Mitchell Award, which was sent to Dr. Lowry. I also sent a copy of the 31-page protest to George Miller, and another copy to the president of the Oregon Chapter of the AMS, who called Miller for advice. The following January that copy was turned over to the next president, TV weather reporter Jim Little, who eventually "lost track" of it. It was never made available to the 40 or so local members.

Dr. Lowry did not respond until 46 days had elapsed, and then only by telephone, and conveying a reluctance to respond in writing. He was disturbed by the evidence in the protest, and discussed ideas to improve future procedures. The national AMS by necessity depends on the professional integrity of the one writing up the nomination, and endorsements of upper level managers.

Lowry claimed that, in this case, the nomination was unusually well-written. My curiosity rose. I asked for identity of the nominator.

Lowry hesitated. That might be confidential. He didn't know for sure, the AMS had rigorous rules, he didn't want to break any rules. But he would find out, and if he didn't call back, that meant it was confidential.

The long distance phone call was October 11, 1983. He did not call again.

On December 5, I wrote Miller again, demanding identity of nominator for the Mitchell Award, and requesting copy of nomination. Phil Peck called December 8, and photocopy arrived January 6. Critics of the NWS need patience.

The nomination described the exceptional character of the storm. References to forecast performance were weak, or merely hyped assertions. Here are some excerpts, with my insertions bracketed:

> By 8 a.m. Nov. 13 gale warnings [for up to 40 knots] had been hoisted along the entire Oregon Coast by Lead Forecaster Ron Surface. He cautioned that storm warnings may be issued later in day.

> At 1 p.m. Lead Forecaster Ron Surface issued storm warnings for the South and Central Oregon Coast [for 40 to 60 knots, hardly alarming for November] and a high wind warning [for SW Oregon for 45 mph, barely garbage-can weather]. He cautioned that should the storm veer northward along the coast, high wind warnings would be extended.

Peck made no further mention of public forecasts by Surface.

At 3:05 p.m., Surface predicted winds for Portland at north 15-25 mph. The wind reached 90 in gusts from the south. Meanwhile at 1 p.m., Wantz and Feris had predicted 70 mph, and about 3 p.m., Portland time, Chuck Wiese in Minneapolis had predicted 90 mph. He had called KOIN-TV in Portland with that forecast, but their weather reporter, John Walls, rejected it. (More excerpts from Peck's nomination letter:)

> As the 00Z [4 p.m. PST] data, ship and buoy reports, began to come in, Lead Forecaster Jerry Youngberg extended storm warnings along the entire Oregon Coast and issued High Wind Warnings for the remainder of the interior of Western Oregon. He forecast southerly winds increasing to 40 to 60 mph inland and 40 to 60 Knots along the coast with higher gusts [how much higher?]. Warnings were issued at 5.23 p.m.

> Because of early recognition of this storm's potential on the part of Lead Forecasters Ron Surface and Jerry Youngberg, accurate and effective warnings were issued far in advance allowing time for thousands of people to take actions to minimize damage and reduce loss of life. The timeliness and quality of their warnings were outstanding and can be credited with providing an exceptional meteorological service to business, industry, and the citizens of Oregon.

> Please use "Pacific Northwest Wind Storm Report" November 13-15, 1981 as a reference.

Just think of the preparations people would have taken if they had been given the predictions of Wantz, Feris, and Wiese at 1 p.m. and 3 p.m.

The National Weather Service has been fooling people for long enough. This book is my alarm clock to wake them up. Congress should terminate all field forecasters in WSFOs, and sell off the graphic display equipment which they use to awe visitors.

Newpapers, and radio and TV stations, should get their forecasts from genuine forecasters. We can soon find out who they are. Analyze their accuracy.

The Anonymous Threatening Phone Call to Chuck Wiese

Now we come to one of the dirty tricks in this story. Chuck Wiese, a native of Portland, had been pressured out of town in early 1981. He found a job in Minneapolis working for Mike Fairbourne, former TV meteorologist in Portland, who now runs an independent operation at WCCO-TV. According to Chuck, that operation is perhaps the best one in the country.

In early 1983 Chuck Wiese moved back to the Pacific Northwest, to work for KING-TV in Seattle. Our old friendship was renewed. Wiese learned about the protest of the Mitchell Award, described above. On January 19, 1984, he wrote to the Weather Forecasting and Analysis Committee of the AMS, endorsing the protest, and nominating William Wantz and Charles Feris for a similar Charles D. Mitchell Award for the same 1981 storm.

On January 25, Wiese received an anonymous threatening phone call in Seattle, a voice saying that because of his letter, **"Your job at KING-TV, and your future in television is in jeopardy."**

He had sent copies of his letter (to the AMS) to some of his adversaries in the Pacific Northwest. No one should forget this threatening phone call.

That event started this curmudgeon on a new career. Conversely, George Miller has written: **"I did not believe**, and still don't, that Chuck Weise (sic) received a threatening phone call."

Chuck is bright, but not cunning, and is a poor actor. He has plenty of points against adversaries, and no need to invent another episode.

From central Oregon, I called Jim Little (the newly elected president of the Oregon Chapter AMS), and John Walls (a known adversary of Chuck), and George Miller and Phil Peck (jointly).

Miller set up a special meeting at the WSFO for February 6, 1984, to "clear the air," but it got even more polluted.

In my protest of the Mitchell Award mailed August 26, 1983, I had no knowledge of Peck's nomination letter or the whitewashed Report of the Pacific Northwest Windstorm (which I obtained February 6 during the special meeting). Here is an excerpt from a tape recording made at that meeting:

Lynott   George is engaged in coverup.
Miller   Come on, Bob, I'm sick. Anything you want pertaining to
         that storm is yours. It has not been that hard to get. I
         assumed everybody has a copy of this report. You can
         have **that one** if you want it. It might be our last one, but

for God's sake take it, and read it, and memorize it. I am
not covering up one thing whatsoever, Bob. I don't play
silly [bleep] games like that.

Lynott   The record shows that you did.

Miller   I'm very disturbed, Bob, that you keep telling me I am
withholding information from you. What have I got to
cover up?

This exchange illustrates the gulf between opposing viewpoints in the coming shakeup. The insights gained at the
special meeting appalled me. I would alert the national AMS
to the problems in the Pacific Northwest.

Correspondence with National AMS

On February 23, 1984 I wrote 6 pages to Dr. Alan I.
Weinstein, the new Chairman of the Committee on Weather
Forecasting and Analysis, enclosing extensive background
material.

I endorsed the nomination of Wantz and Feris made by
Wiese, and refuted the nomination from Phil Peck, and
described my alarm at the harassment of Chuck Wiese, and
the role of the NWS in the decay of forecasting.

Dr. Weinstein answered on stationery of the Department
of Navy, Office of Naval Research, his employer.

We need more people with strong opinions who are
willing to stand up for what they believe in . . . The award to
Surface and Youngberg was made in good conscience based on a nomination that the committee felt justified the
award. I won't debate the wisdom of the award. I will defend
to the end, however, the good intentions of all the committee members and the nominators.

What is done is done! There is nothing to be gained in
further airing of the past. Rather, what can we do about the
future? . . . In closing, let me reiterate my praise for your
commitment to what you believe is right. Please don't make
this commitment appear as some kind of vendetta.

The good intentions of the committee members were
modified by the need to further the best interests of their
agency or their institution.

I replied, as follows:

My protest of the award did not imply any rescinding.
My concern is now focussed on a larger problem.

Unawareness, indifference, and distortion prevail, both in the AMS and in the public. My intent was to funnel material to your committee members. They could integrate ideas from various sources and forward them to the Council.

I'm sorry you suspect my motive is some kind of vendetta. If something is broke, fix it. Everything except the instantaneous present lies in the past. The future is unknown. We can improve only if we examine past errors. Much is to be gained by airing the past, not only the recent past, but anything still pertinent. I didn't mention anything more than 35 years ago. Everything since World War II pertaining to forecasting is related to the present deplorable state of affairs.

My target is the lack of genuine weather forecasting, the subtle substitution of nowcasting and news reporting, the deliberate blurring of the distinction, and the accompanying bureaucratic resistance to competition.

The sacred cows of **public confusion** and **protection of life and property** will lose their aura of reverence when the public realizes they receive no **advance** warnings (with minor exceptions) of significant weather changes.

Your committee may choose to remain aloof. It would seem your committee ought to be concerned. A transition is shaping up.

Forecasting is much more difficult, and more risky, than reporting.

To obscure its lack of skill at forecasting, the NWS strives to blur the difference between reporting and forecasting.

The NWS defines a "warning" as a notice of dangerous weather already occurring.

5

Chuck Wiese, Troublemaker

The National Weather Service is striving to maintain its monopoly on public weather forecasting. One effort is by effusive propaganda. See how much we do! Our product is outstanding! We have all this modern equipment, and man power, and scientific skill! Please increase the appropriation so we can serve more.

Another effort is undercover, to eliminate any competition which might arise. This chapter is about a young man who has been harassed because he dares to compete with the NWS on forecast accuracy. He strives to be an independent forecaster.

Chuck Wiese has been branded by the regional NWS as a troublemaker. I agree but with a different meaning. Chuck's adversaries claim he alarms and confuses the public. I only note he usually scoops the federal forecasters when significant changes occur. He really does make trouble for them.

When the federal monopoly is broken, and genuine forecasters like Chuck Wiese take over responsibility for public weather forecasting, the Promise will be kept for the Weather Tomorrow.

Chuck Wiese, age 30 in 1986, is a television forecaster in Seattle, Washington. I have unexpectedly become his mentor. The hostility of the NWS "effectively" ran Chuck out of Portland by hindering his employment as an independent forecaster. Thus the access of the local public to better forecasts has been hindered. The reader will be surprised at what goes on behind the scenes.

In the art of local weather forecasting, skill is accumulated by reviewing episodes involving significant changes. The case studies, or case histories, in this book are slightly different in nature. But they are related to factors hindering forecasting skill. Although these episodes took place in the Pacific Northwest, they likely will provide road maps for progress in any locality.

The Early Years of Chuck Wiese

Chuck Wiese was born in Portland April 24, 1956. At age 6 he was thrown to the ground by the wind during the Columbus Day storm. Few careers begin with such stimulation. He started to watch Capell and Lynott on television. Learning to read, he visited the library for books on weather. His parents gave him a backyard weather station for Christmas.

The glow of an entrepreneur began to show. At age 11 he was presenting daily weather reports on KBPS, the public school radio station. Moving up, it was KEX commercial radio at age 12, and KOIN radio at age 13, reaching the heady salary of $5 each weekday afternoon, equivalent to $17 in 1987 dollars.

Wiese was too young to drive. His mother had to chauffeur him to the Weather Bureau office at the airport to gather data and study weather maps. His attitude was not sufficiently subservient to please some of the Bureau personnel, who regarded him as a nuisance, or a brash upstart amateur intruding on professional turf. Occasionally they would hide his briefcase, or complain about their unwanted role of baby sitter. One said his mother had made a mistake in having him. The adversarial relationship had an early start.

In September 1971, at age 15, he was hired for the 5 p.m. Saturday weather show on KOIN-TV. (I had left TV in April 1968) Chuck frequently called Jack Capell for advice. A few months later the 11 p.m. weathercaster was fired, and Chuck added the late evening weekday shows to his schedule, meanwhile continuing school and the afternoon radio shows.

The workload was heavy, and the 11th Grade was boring. Discovering that a GED test would substitute for a high school diploma for college entrance, he passed the test with a high score, and dropped out of school in November 1972, age 16. In December he appeared on the "To Tell the Truth" television show in New York City, a trip with all expenses paid. That was his first trip on a jet airliner.

Now he wanted to be an airline pilot. No longer in school, and with income to pay for flight lessons, he earned a private pilot's license in April, just before his 17th birthday.

His career suffered a temporary setback. A new director of TV News, Ted Bryant, hired Rick Elgin for the 5 p.m. and 11 p.m. weekday weather shows, and cut Wiese to only 4 shows on weekends. The weekend shows had expanded to 5 p.m. and 11 p.m. on Saturday and Sunday. This was not enough income for Wiese, who left in August 1973 for a TV weatherman job in Ft. Wayne, Indiana. He was 17.

Promises by that station to install minimum weather facilities were not kept, and Wiese returned to Portland in January 1974, getting back the 4 weekend shows at KOIN-TV. That assignment temporarily increased to 7 days a week, 14 shows, during the summer of 1974 because Rick Elgin left KOIN-TV.

Chuck needed a college degree, and took leave in September to enroll at Oregon State University at Corvallis to study meteorology (naturally). John Walls was hired as No. 1 Weatherman, where he remained for nearly 12 years. No one predicted the storm ahead.

Before the end of his first year in college the mutual needs of Wiese and KOIN-TV got together again. He was rehired for the 4 weekend shows (even though it meant weekly commuting) a position that would continue for 6 years.

A career in aviation still beckoned. While studying at Covallis he intermittently continued flying lessons, acquiring an instrument rating in 1975 and a commercial license in 1977. The commuting to Portland for 2 days each week was a constant burden.

In June 1978, target date for graduation, Chuck lacked 7 quarter-hours of credits, which he completed in December. Because commencement ceremonies are only held once each year, the school issued a letter confirming completion of the cirriculum, and awarded the diploma in June 1979.

The academic credits required at a university for a bachelor's degree in meteorology constitute the generally accepted "union card" for a professional meteorologist, but there are many exceptions.

Some short-term lower-grade officers in the military, such as ROTC graduates, with a degree in another field, are given limited courses in meteorology, and are accepted as professional meteorologists. Also, the recent propaganda by the NWS to label TV weather reporters as private professional meteorologists has blurred the definition.

The standard bachelor's degree in meteorology includes general basic physics (much of which is unrelated to weather forecasting), mathematics and statistics (overrated subjects, 70% of which are a waste of time for a forecaster), and assorted courses in climatology, atmospheric physics (important), physical geography (very important, but often left to self-study), and synoptic meteorology (most important of all, but seldom emphasized).

Few administrators in academia understand weather forecasting. They are unlikely to gain much insight from perusing the scientific publications in the comparatively narrow field of meteorology. The few skilled forecasters are not prolific writers for scientific journals.

Actually, the main requirement for skill in forecasting is an Alan Kent second-rate mind (often described as "aptitude", but properly defined as analytical ability), and an opportunity for several years of learning-by-doing. Such opportunity is uncommon. It is small wonder that good forecasters are scarce.

The usual course in synoptic meteorology provides less insight than one gains from personally forecasting every day for six months. By December 1978 Chuck had accumulated 7 years of part-time and full-time forecasting, making his own forecasts, and had studied weather maps for several years before that.

Like many forecasters in earlier days, Chuck first acquired operational skill from hands-on experience, then later completed the formal studies presumed necessary. Ordinarily, a new graduate faces several years of internship. Chuck already had completed that requirement.

Although he was fully "prepared," his television forecasting remained limited to weekends only. Adding a multi-engine rating to his commercial pilot's license, he found intermittent jobs piloting corporate aircraft. But the chance to become an airline pilot now seemed dim.

Wiese knew his TV ratings were high, which translated to advertising dollars for KOIN-TV. His occasional scoops over NWS forecasts attracted favorable attention, except from a few. Naive and self-confident, he believed competence in forecasting would lead to success. KOIN-TV's John Walls, saw things differently.

Part of the Pattern of Silence

The story of the adversarial relationship between John Walls and Chuck Wiese is handicapped by silence from Walls. This is a matter of public interest because it goes beyond personal animosities. It involves public weather forecasting. My repeated requests to Walls for information have been spurned. The story must partly depend on indirect evidence.

In 1979 John Walls was established as the Chief Meteorologist at KOIN-TV. The time-consuming trips to the airport for data had long since ended. The station now had a fully equipped weather office. With a $100,000 loan from Lee Enterprises, the major owner of KOIN-TV, Walls had organized a "private" forecasting company, called Lee Weather Northwest, composed of several partner meteorologists, including newly graduated Phil Volker, who had ambitions for television.

The company was set up under the monopolistic division of turf which the National Weather Service is striving to elevate from "customary" to "traditional" without benefit of congressional authorization. Walls decreed the policy for KOIN-TV: Only NWS forecasts would be disseminated to the public. Independent forecasts were permitted for private clients only. My request to see this station policy was ignored.

KOIN-TV had pioneered with its weather program, "Mr. Weatherman," for 14 years, from late 1953 to early 1968. Chuck Wiese was continuing with a similar independent forecasting attitude, and with success, even though appearances were limited to weekends, or were intermittent, because of schooling. However, the scoops of Wiese were embarrassing to the NWS.

The reader might ponder a policy which permits independent forecasts for private clients only, and NWS forecasts for the public always. This would seem to imply the independent forecasts are superior, because a fee is charged. The NWS forecasts are free (to the individual, but taxpayers pay for them). It seems to follow that the public would be short-changed.

This puzzle is never answered by the NWS. They strive to define "specialized forecasts" as forecasts for weather elements which have no public interest, such as the minimum temperature tomorrow morning in a certain orchard.

But many "private clients" of independent forecasters have the same interests as much of the public--will it snow, how much, will the wind blow, how strong, and how cold or how hot will it be?

Chuck Wiese ignored the policy of Walls to use NWS forecasts only, because it was beneath his integrity as a forecaster. Discussions with Walls were fruitless. Each time Wiese issued a forecast different from the NWS, somebody from the WSFO would phone management and complain, even though all, or nearly all, of the independent forecasts were correct.

When a midday news show was added to local programs, the weather segment was assigned to Phil Volker. Wiese was low man on the weather totem pole.

When Phil Volker told investigative writer Robin Cody: "He (Wiese) was terrible. He was wrong all the time. Always alarming people.", Phil was merely repeating, like a broken record, the established catechism of the Portland NWS.

On July 28, 1985, when the Cody story was printed, Volker tried more than 20 times to reach Wiese in Seattle, to deny the quotation. Volker also denied it via a letter to the editor.

Chuck prefers to let the tribunal of public opinion settle these matters, rather than a court of law.

To supplement his weekend TV job, Wiese found a part-time job in September 1980, forecasting weather for KXL radio on weekdays. Access to weather data, denied on weekdays at Lee Weather Northwest, required him to visit the fireweather office at the Forest Service, or the FAA office at Hillsboro airport (which served pilots). Even on weekends for the KOIN-TV shows, Walls hindered Chuck's use of weather maps and teletype reports.

Walls complained to management that Chuck's independent predictions on weekends (without regard to their accuracy) were disrupting the relationship between KOIN-TV and the National Weather Service. Such would reduce that agency's "cooperation" in furnishing official forecasts, warnings, and other "products." (This is one way the NWS exerts pressure behind the scenes.)

On Saturday, January 17, 1981, Chuck Wiese alerted viewers to the existing snowpack in the mountains, especially in lower elevations, and the approaching warm rain. Noting the similarity to the conditions for flooding in January 1965 he explained the situation warranted close attention. The official forecasts had not yet suggested any problem.

Videotapes of news programs are routinely kept for one week. Anticipating harassment, Chuck made a separate tape of his program. John Walls complained to management that Wiese had issued an unofficial flood warning.

James D. Wakefield was Meteorologist-in-Charge and Area Manager at the Portland WSFO. He called Station Manager Howard Kennedy, claiming that river forecasting personnel of the NWS saw Wiese on the air, assumed a flood was shaping up, and came to work voluntarily Sunday morning. Wakefield supposedly was forced to pay unauthorized overtime.

The station videotape had been erased, but Wiese produced his own for KOIN-TV management. No flood warning had been issued by Wiese.

I found no evidence at the River Forecast Center, which is administratively independent of the WSFO, to support the claims of Walls and Wakefield.

Unknown to James Wakefield and John Walls, Chuck Wiese was already preparing to leave Portland. Disparaging rumors about Chuck had circulated to other television stations that might have offered employment, even to a small independent forecasting company, Intrawest Weather, in Bend, Oregon. "A prophet is never honored in his own country."

In the television column "Behind the Mike" in *The Oregonian* for March 18, 1981, Peter Farrell wrote:

> Portland's one-time schoolboy weather forecaster is about to go to Minneapolis and join a full-scale weather war. There he will work full time on an expanding weather operation headed by Mike Fairbourne, who went to WCCO-TV from KGW-TV in Portland in 1976.
>
> [Chuck] Wiese said he read in a trade magazine about plans at WCCO to outdo the massive weather effort of its chief rival, KSTP [which had 9 meteorologists]. "WCCO, TV and Radio, plan to hire four or five more meteorologists and to spend millions to beat KSTP. They plan to develop a weather forecasting center that will be completely independent of the National Weather Service.."

Weather is important to citizens of Minnesota. They want to know the Weather for Tomorrow, and they compare the forecast with what happens. Public weather forecasting in

Minneapolis is dominated by two independent and competitive forecast offices, not the NWS. The people get genuine forecasts.

Chuck sent Fairbourne a videotape and a résumé. He flew to Minneapolis for an interview, and a job offer followed.

Violent thunderstorms and occasional tornadoes are among the weather problems in Minnesota. Chuck asked Fred Decker, his former college professor at Corvallis, to come to Portland and show him how to operate radar equipment at the WSFO at the airport. Wakefield granted permission.

Decker arrived first and was in conversation with Wakefield when Wiese arrived, and sat down outside the open door to Wakefield's office. Wakefield was complaining about independent forecasting by Wiese, and waved a letter he was sending to KOIN-TV management, which Howard Kennedy had requested after Wakefield's telephone complaint about the so-called flood warning by Chuck. Wakefield told Decker if the NWS had someone like Chuck they would fire him.

Similar sentiments presumably were in the letter, which has not yet been found, but it may someday. Wakefield has refused to answer my questions about this episode. Chuck was outraged, demanded apology, and threatened legal action. However, the move to Minneapolis discouraged that effort.

The Walls-Wiese controversy did not renew until the special meeting at the Portland WSFO on February 3, 1984. This time Chuck had a kindred spirit for an ally, a "curmudgeonly old crank."

The main requirement for skill in forecasting is an Alan Kent second-rate mind (also defined as analytical ability).

Wakefield told Decker if the NWS had someone like Chuck Wiese they would fire him.

6

The Famous "Dr. John"

In headlines 3/8 inch high in *The Oregonian*, January 25, 1986, Peter Farrell (Behind the Mike) wrote:

Dr. John to quit KOIN weather scene

Dr. John Walls will do his last weather forecasts for KOIN (6) on February 28 after nearly 12 years with the station. Walls said he is leaving KOIN for personal reasons not connected with his health, and news director Ted Bryant said Walls "just decided he didn't want to do television anymore."

On March 5, 1986, Farrell published u smaller item:

Miles Muzio joins KOIN-TV March 31 to do the weather on the 5 and 11 p.m. newscasts. He replaces John Walls, the famous "Dr. John," who left the station after a disagreement with management.

Yes, John Walls is famous. How did that happen? He was Chief Meteorologist on the TV station with the highest ratings in Portland for news. The Vice-President and General Manager, Mick Schafbuch, assured me (April 20, 1984): "We are very pleased with John Walls, his ratings and his professionalism."

I have been investigating the story for 3 years. Much remains hidden.

My evaluation is limited to the subject of weather forecasting and related activities. The contradictions in statements by John Walls are puzzling because they are so obvious.

The story of John Walls is important only because it reveals the magnitude of the NWS stranglehold on weather forecasting. Walls identifies himself as Dr. John Walls, Meteorologist. His personalized auto license reads "Dr. John." He is proud of his academic credentials. On TV he presented "his" forecasts.

When Robin Cody explained the nature of the PhD displayed by John Walls, John wrote to the editor of *The Oregonian* (printed September 1, 1985):

> I am very unhappy with the extremely distorted picture of my background in Robin Cody's July 28 cover story, "Why Can't They Get It Right?" It is true that my doctorate is in Adult Education and included a dissertation dealing with the production of visuals for adults. But that is precisely my job at KOIN.
>
> I believe that my degree is extremely applicable to my position. I also hold a master's degree in meteorology (completely ignored by Cody). This degree was received from the University of Washington, and my initial graduate work was at the University of Chicago, the same institution where Bob Lynott claims training.
>
> I have 34 years experience in weather forecasting as a qualified meteorologist. Twelve of those years were in Portland. Five were at the largest known weather facility in the world, the Air Force Global Weather Central. These facts also were completely ignored by Cody.

I will not ignore any facts about Walls. The problem is to uncover them.

His letter-to-the-editor illustrates the gulf in communications in the meteorological community. He said "34 years experience in weather forecasting as a qualified meteorologist."

To most people this infers not only that a forecaster should be a qualified meteorologist (and Walls probably meant one with one or more degrees), but also that a qualified meteorologist is also a forecaster.

Actually, few meteorologists are "qualified forecasters." How can a meteorologist who merely, and only, disseminates NWS forecasts call himself a forecaster? A real forecaster is a decision-maker, not just a disseminator.

Let's begin with John's academic credentials.

John Walls retired as a Lieutenant Colonel from the U. S. Air Force in 1972. He earned a BS degree from Butler University in Indiana in 1952, an MS in meteorology and climatology from University of Washington in 1961, an MS in education (Counseling and Guidance Secondary) from University of Nebraska (Omaha) in 1971, and a PhD in adult education from University of Nebraska (Lincoln) in 1974. His PhD thesis was "Visual Materials for Individual Counseling of Adults in a Community College."

Walls posed as having a doctorate in meteorology. When someone is introduced on TV as "Dr. John Walls, Meteorologist," most people assume he has a PhD in meteorology. Convention implies that. He was not introduced as Dr. John Walls, visual aid specialist, with an MS in meteorology.

I am careful not to ignore the master's degree in meteorology from the University of Washington. Upon inquiry, the Registrar would only confirm that Walls had been granted such degree, and declined to answer a question about a possible thesis. I checked further.

John's thesis has library number 551.5, Th 11537, entitled "Meteorological Education in the United States," which was partial fulfillment of the advanced degree in meteorology. It had only been checked out once, in 1964. The study was an analysis of answers to a questionnaire sent to various educational and scientific organizations in the nation and abroad, to determine what was being taught at that time.

The bibliography filled 42 pages, and included catalogs for weather instruments, tables of sunrise and sunset, and the booklet for the Boy Scout merit badge in weather. Walls ignored my request for information about any other achievements by him for this degree.

John claims he had 12 years experience in weather forecasting at KOIN-TV, where he was Chief Meteorologist. But he also admits that the production of visuals was "precisely" his job there. An article by Julie Bookman in *The Columbian* (Vancouver WA) January 14, 1985, quoted John:

> In essence, I put together a slide show every night. Information for each weather segment is actually stored electronically on computer disks.
>
> Most people believe we actually make the forecasts. We don't. We work very closely with the NWS and then report the information provided.
>
> I somewhat fell into the profession during the Korean War when the military was desperately in need of weather people and there I was — with a math background.

Ms. Bookman said Walls prides himself on being a meteorologist rather than a weatherman who is more of a television personality.

It is remarkable how persuasive advertising can be. An

advertisement in *The Oregonian* showed a young lady's picture next to a picture of Dr. John. She is quoted: "The difference he makes and the reason I watch Dr. John is his background . . . I trust his judgment!" In television, fame often is a product of the promotion department.

Let's preserve some of Dr. Walls' philosophy about forecasting.

Phone call to John Walls

On January 30, 1984, after the anonymous threatening phone call to Chuck Wiese, I phoned John Walls: (slightly condensed, my insertions within brackets)

Lynott I have some pointed questions to ask you. I'm making a recording so there's no question about accuracy of what I remember. (Walls: Okay.) Are you aware that Chuck Wiese had a threatening phone call?

Walls No, I'm not.

Lynott This is kind of serious. It was anonymous. [I reviewed the secrecy concerning identity of Phil Peck, nominator] It is obvious that Surface and Youngberg did not do the best forecasting that day, of anybody.

Walls That's where I disagree with you.

Lynott I have one more question. On your program there's quite a bit made of the fact that your weather forecasts are approved by the AMS.

Walls No, not approved.

Lynott That's the word they used.

Walls They did, about the first 2 or 3 days. But I made them change it. I objected violently to the word "approved."

Lynott That's not quite the way it is. (Walls: No, it's not.) They have endorsed your weathercasting. They say it meets standards. Now here's another thing, you are advertised as being a Doctor.

Walls Yes. It's in adult education with an emphasis on visual preparation, and I think it's a perfect combination with a meteorology Master's. I have two Master's degrees, and a Bachelor's. The emphasis all the way through is in visual preparation, in the Doctorate.

Lynott Do you consider yourself primarily a forecaster?

Walls For purpose of television, no, I do not. In fact, probably Chuck Wiese has mentioned [John's policy], anybody doing weather in television should **not** advertise themselves as doing forecasting. That's one of my strong philosophies.

Lynott Why do you object to competition in public forecasting?

Walls   Oh, I don't at all.

Lynott If weathercasters are supposed to do what the NWS says, that looks like It supresses competition.

Walls   Wait a minute, I said the forecasters on TV should be in the primary job of **interpreting** as accurately as possible, using the most **attractive** visual aids as possible, in the most **understandable** way as possible, the NWS product.

Lynott Why should it be limited to that?

Walls   Because I don't feel, and this is one of my biggest objections to Chuck Wiese, that a person in any kind of a forecasting job, when they should be spending 90% of their time developing visuals, can use [the remaining] 10% of the time to do an **accurate** forecasting job. If they are doing their job on television developing visuals, then they don't have time to do forecasting. That's my philosophy.

Lynott I appreciate your open opinions and candor.

Walls   The only circumstance where I can see where a television forecaster can do other than that, is when the TV station is co-located with a consulting firm where they have the staff, to watch weather 24 hours a day. [Then they] can out-forecast, if you want to put it that way, what the NWS is doing.

         I've watched Chuck Wiese time, and time, and time again, walk in to that TV station down there, and in 10 minutes has the whole thing figured out, and forecast something completely different from what the NWS was saying. And that was one of my biggest objections to Chuck.

         And then I had to follow that up. To me, that is really unethical, to say that he can do that. I've been in this business for 30 years. I've never seen even some of the most outstanding forecasters in the country be able to do that.

Lynott If somebody does something wrong, don't you think the market place will catch up with him, and he will lose his credibility? If his forecasts are inaccurate, don't you think that will rapidly catch up with him?

Walls   As far as the weather goes. [John changes the subject] Look at Willard Scott. He's got a lot of people that watch him. I don't think they watch him so much for the credibility of his weather forecasts. They watch him to be entertained, as a whole lot of people do, when they watch TV weather.

Lynott   What is your view about someone who primarily wants to put out the best possible forecast, without putting primary emphasis on the graphic aids? He wants to emphasize the quality of the [forecast] product.

Walls   That's the ultimate goal.   I'm saying one person can't do that, I don't care who they are.

Lynott   That's an interesting point.

Walls   Just like a sportscaster, for example. He could try to predict the outcome of every college football game in the country, if he had a staff of people providing information about all those teams, to do it with any accuracy. It's exactly the same thing with weather. You have to have assistants, you can't do it by yourself. Unless you got a big consulting team behind you, you **have** to depend on the NWS. They are the ones that are in the business. They got the staff.

         Most TV stations only have **one** weather guy on the staff. How in the world can they spend any time at all in the development of something to put on the air without having some information collected and looked at for them. They just can't do it by themselves. Most TV stations don't have the communications close to what they need. They certainly can't walk in, and in 10 or 15 minutes say, hey, this is going to happen, based on intuition, which most of it is. I don't know of anybody that can do that.

Lynott   I don't think anybody can do it in 10 or 15 minutes either. If anybody tried to do it in that short of time, without adequate preparation and analysis, and also the factor of competence enters into it--

Walls   As I say, Willard Scott, I enjoy watching him to be enter- but I don't care for his product at all. It depends on the purpose of what you want to do. Jim Bosley, I got a lot of respect for Jim Bosley [a rival on KATU-TV] as an entertainer, as a talker I think he is great, but I wouldn't give a nickel for what he knows about the weather. I think I know something about the weather, but my TV presence, I think it's pretty bad, as a communicator. Unfortunately, one is a science, and one is an art, and you find very very few people trained in both sides.

Lynott   You have answered my questions. I have better insight into your feelings. I appreciate your comments.

Walls   Okay, Bob.

Lynott   Thanks a lot.

This conversation is revealing. John believes a weathercaster should not advertise himself as doing forecasting (but he claimed 12 years of experience in forecasting on TV in Portland). A weathercaster should spend 90% of his time developing visuals. To out-forecast the NWS one must watch the weather 24 hours a day.

One person alone can't put out the best possible forecast. Unless one has assistants, a big consulting team behind you, you have to depend on the NWS.

It is unethical to forecast something completely different from the NWS, especially if done in 10 minutes, which means the forecast is based on intuition.

John Walls claims Jim Bosley approaches weathercasting as a form of entertainment, like Willard Scott.

John claims he is on the side of science. Let's see how science and John Walls did on Friday, November 13, 1981. We already know how the NWS did. They got the Charles D. Mitchell Award.

How John Walls Missed His Chance to Be a Hero

On that afternoon the NOAA weather wire (the statewide teletype circuit carrying the "products" of the Portland WSFO) made no mention of the alarm issued to electric utility organizations by BPA (Wantz and Feris).

Chuck Wiese, at WCCO-TV in Minneapolis got excited, (but then he is wrong all the time, always alarming people.) He phoned KOIN-TV in Portland, and talked with Ken Woo, the news anchor preparing for the 5 p.m. show. Excerpts of a taped interview with Woo follow:

> That Friday Chuck called me about 1 p.m. that we were going to have some powerful winds. It might be as powerful as the 1962 Columbus Day storm. I went over to the news wires. I didn't spot anything. I went in to our Assistant News Director, Craig Kuhlman, and informed him of what Chuck had told me.
>
> John Walls came into the office, and he was extremely upset about what I was telling Craig. He got extremely angry, started shouting that Chuck was irresponsible, that the storm was going to miss us, and I shouldn't pay any attention to what Chuck was saying.
>
> I said "But John, as far as my experience with Chuck is concerned, he has been right more times than he has been wrong, at least on the big things that matter." John

stalked out of the office, extremely angry. I alerted the crews that come in early in the morning. I consulted with Phil Volker, the second meteorologist at KOIN. "Yeah, it could happen, then again it may not."

Chuck called me back around 4 p.m., updating me. I went over to John Walls. He went into a big tirade: "I don't believe any of it. You are wasting your time. Nothing is going to happen."

At 5 o'clock we were all on the air, three stations. Channels 2 and 8 led with their meteorologists, that this was a very serious storm. -John downplayed the whole thing. When he came on at 5:20, he didn't even mention we had a potential problem.

What happened later that evening, it's just hearsay. I was not there. I was told that at 11 o'clock he again downplayed the whole thing.

In the years I was at Channel 6, I know that John and Chuck had an adversarial relationship. John had several times gone to management, and in at least one instance, management had told me to watch Chuck, and keep a lid on him. My reaction was--it is not my job, I'm not his supervisor. As to what his forecasting is, I am not qualified to make those judgments. I have too many things to do, too many other responsibilities, and I am not his babysitter. He has always been professional, and he has been right more times than he has been wrong. With that kind of track record, I am not going to quibble with him.

John Walls had complained that one of his biggest objections to Chuck Wiese was the necessity "to follow up" after those instances where Chuck made predictions much different than the NWS.

We don't know what John meant by "follow up," because the responsibility for the accuracy of a forecast rests only on the originator. But as Colonel Ellison revealed, a bureaucratic mind is preoccupied with authority and control, and is eager to assume responsibility for another, if that also means control.

We can guess that in the case of Wiese versus Walls, the usual problem facing Walls was answering the question from others--How come Chuck scooped all you other guys? Yes, Chuck deserves his label of **troublemaker**.

The viewers in Minneapolis were pleased that their independent forecaster had scooped the Mitchell Award boys in

Oregon. WCCO sent Chuck to Portland the next day to get copies of TV videotapes of storm damage.

John Walls' Documentation of "Unethical" Forecasts by Chuck Wiese

On January 30, 1984, after the anonymous threatening phone call to Chuck Wiese, I also called George Miller and Phil Peck at the Portland WSFO. I told them about the call, and complained about secrecy surrounding the Mitchell Award, and the stalling tactics of Miller.

Miller made excuse he did not identify Peck as the nominator because he didn't know Peck was the nominator. He seemed puzzled about my motive in protesting the Mitchell Award.

Miller said, "I get the feeling, Bob, that you still got a [grudge] for the Weather Service, and that kind of bothers me."

The historic meeting of February 6, 1984, at the WSFO (to clear the air), included Miller and Peck of NWS, Charlie Feris of BPA, John Walls, Chuck Wiese (who invited himself), and me.

The 80-minute meeting mostly was about the windstorm of November 1981, and my formal protest to the national AMS. Here are some excerpts from a tape recording of that meeting, with my insertions in brackets:

Peck   [beginning] I don't know why John [Walls] and Chuck [Wiese] are here.

Miller   I asked John to come because of the correspondence he was involved in. I didn't think to call [Chuck]. I presumed you were in Seattle, but you are welcome.

Wiese   [later] That's why I objected to the Award [made by] the AMS. It was not an exceptional forecast . . . It was an under-forecast by the Weather Service. Charlie [Feris] and Bill [Wantz] forecast peak gusts of 70 here in Portland and 90 along the coast.

Walls   I think it is completely irrelevant, the topic. I object strongly, and I've been very critical of you [Wiese] on this, and I have it documented, that in the wintertime, every time there is a low pressure center off the coast, to say on the air, which you have done many, many times, we've got a Columbus Day storm, off the coast. And every time that happens I am flooded with phone calls

	from people alerted up and down the valley with Columbus Day storms.
Peck	So is the NWS.
Wiese	Hold on a second. I am aware of statements you have made to people recently, John. I take exception to some of those.
Walls	I don't think this is the place to discuss it. [Then why did John bring it up?]
Miller	I don't think this needs to be brought in.
Wiese	This will take one minute.
Peck	Let's stick to the topic at hand.

Wiese	[later] I want to say something . . . to get it all out in the open. [Walls had brought it up earlier] I want everybody to hear this. John, you had a conversation last week with Bob, right? And you told Bob that your biggest objection to my working with you [past history at KOIN-TV, prior to 3 years ago] was that I can come in, and in 10 minutes look at a weather map--
Walls	Yes, I've seen you do that more than once.
Wiese	— and come up with a forecast completely different from the NWS?
Walls	Yes, I fully support that. I've seen you do it time after time again.
Wiese	You also call this unethical? (Walls: Yes) I'm asking you to put this in writing. [Which this author is doing right now]
Walls	Why should I?
Wiese	Because you're saying it to these people, and you are saying it to other professionals —
Walls	I agree with it. I was your supervisor, and my —
Wiese	I want this out in the open. You have made these contentions. I know you can't prove them because they are false.
Walls	I got it in writing. I've written it, and it's in your file [at KOIN]. It's still there.
Wiese	I'm telling you that unless you're willing to write a formal complaint and submit it to the AMS —
Walls	I'm not going to do that! It has nothing to do with the AMS!
Wiese	It does! Because you're telling it to other people.
Walls	Chuck, I'm going to do what I want to do.
Peck	I think we've had enough.

Later I reviewed the tape, and wrote to John Walls:

Dear John: In our phone conversation you expressed

criticism of Chuck Wiese regarding over-forecasting of severe weather. During the February meeting you brought it up again, saying you have your criticism documented. You said your objections to Chuck's conduct were in writing, in his file at KOIN-TV presumably.

I have an open mind. Maybe Chuck's conduct has been out of line, and I would like to review your documented criticism. May I see the details? I am trying to understand the background behind your conversations. Maybe you can convince me.

Walls did not reply. I phoned him from my home in central Oregon. The matter was at the bottom of his priorities. He had thrown away my letter.

Chuck Wiese obtained permission from the news director to drop by and inspect the "evidence" in his personnel file. Chuck was kept waiting for about and hour. There were only a few innocent pages in the file.

On April 17, 1984 Chuck Wiese and I went to the WSFO, armed with a copy of the Freedom of Information Act, and the Privacy Act.

The Privacy Act allows an individual to review almost all Federal files pertaining to himself. It states that secrecy in government is "the incubator for corruption." Agencies must keep an accurate accounting of all disclosures made to other persons. Moreover, this information must be maintained for at least five years.

It also states: "The privacy of an individual is directly affected by the collection, maintenance, use, and dissemination of personal information . . . Opportunities for an individual to secure employment . . . are endangered by the misuse of certain information systems."

We asked to see a copy of the letter which James Wakefield had waved in front of Dr. Fred Decker in early 1981, the one he was intending to mail to KOIN-TV complaining about the unofficial flood warning by Wiese.

While waiting an hour for George Miller to arrive, I noted Audrey Kellogg, stenographer, was still working there. I remembered her as Wakefield's long-time secretary, a gentle, courteous lady.

In civil tones we explained our mission. We had no tape recorder. Such precaution had not entered our minds. But we realize now that would have been prudent.

Was there anything concerning Wiese in any files? Miller

replied negatively, but offered free opportunity for search. Miller's word was quietly accepted. We had no desire to intrude in Miller's files. We just wanted that letter.

I politely asked Kellogg specifically about the sought-for letter. She disclaimed any knowledge. Miller explained office files had been pruned of out-of-date material a few months earlier, and they were not required to keep files more than two years, except forecasts for five years.

I asked Miller for copy of station roster, unrelated to search for Wakefield's letter. I had not been to WSFO for several years, and wanted to get up-to-date on present personnel. Copy of roster was supplied. Although disappointed about failure to obtain Wakefield's letter, we maintained decorum.

Then Wiese and I went to KOIN-TV downtown, to take up the matter of the "documentary evidence" of the unethical forecasts by Wiese.

After long waiting in the lobby for the news director, Chuck asked to see the general manager, Mick Schafbuch. Although we lacked an appointment, Schafbuch came out to lobby. Chuck explained he did not want his nine years of creditable service at KOIN-TV to be distorted, and used to hinder his career. Schafbuch absented himself for less than a minute and ordered personnel manager, Bob Burke, to locate Chuck's file.

Then he invited Wiese and me to accompany him to the personnel office to review Chuck's file. We met Burke in hallway, on his way to boss's office. But I did not know Burke, and was puzzled why this man happened to be carrying Chuck's file. The thin manila folder contained only about 6 unimportant pages. Had anything been removed? No.

Later I recalled a well-filled large brown envelope also in Burke's hand, but opportunity was not grasped by the forelock. (I should have said, "What else do you have in your hand, Mr. Burke?")

John Walls' endorsement of the Mitchell Award to Portland forecasters

During the discussion about the windstorm of November 13, 1981 at the hard-knuckle meeting on February 6, 1984 at the WSFO, John Walls voiced his support of the AMS Charles D. Mitchell Award to the Portland forecasters. From the tape recording (condensed, my insertions in brackets):

Walls   The original complaint [this author's] was the Award from the AMS, right? It has nothing to do in that sense [with us here]. Anybody could write a recommendation for an award. It is up to the AMS to make a decision, based on the nomination, whether to award that particular award. I don't see where it is anybody else's business, except possibly the local AMS Chapter. And I don't see why any of us at this table are concerned about it.

Wiese   We don't feel the Award was given to the NWS properly.

Walls   Then talk to the AMS! The NWS didn't make that Award. I don't see how you can complain to these people, because the AMS made the Award. I don't see any reason for discussing this.

Walls   [after a detour in discussion] When you write a recommendation for an award, does a copy go up through your channels, to the Region? Did you get any kind of a comeback from the Regional Office, that had anything contradictory to say about it? (Peck: No) If the AMS had nothing to say, and the NWS had nothing to say, I cannot see why it's anybody else's business!

Wiese   John, we're making it our business.

Walls   Then don't come to us. Go to AMS. I can see no reason to get into the technical aspects of it, the forecasting ability or anything else. I don't see where it's anybody else's business. If you want to write, write to the AMS. It's their award. If you're challenging it, they'll tell you why they awarded it. Why sit here and waste everybody else's time talking about something that is absolutely none of your business.

Wiese   It is our business, that's why we're here.

Walls   These people didn't give that award.

Wiese   [after another detour in discussion] John, you support the award to the NWS, right?

Walls   Very definitely.

Wiese   I think you ought to put that in writing.

Walls   I don't see any reason for that. If I had been asked to write the nomination, I would have. I think its irrelevant.

Wiese   You ought to defend the WS. You are siding with them.

Walls   If they want me to, I will. It has been well supported by the NWS, and the AMS. I see no reason to go further.

The NWS promotes media weathercasters as part of the public-private partnership. They are loyal supporters. Note how John Walls said, "Don't come to **us**".

Tape recorders are useful in meeting Werner Baum's call for a more penetrating analysis of the strengths and weaknesses of the actors in this drama.

Any serious forecast user (SFU) must ponder the question: why should John Walls, the Chief Meteorologist at KOIN-TV, oppose the independent forecasting done by Chuck Wiese on weekends? If **any** weather forecaster on TV establishes a reputation as an unjustified alarmist, or a forecaster more inaccurate than the NWS, the TV station will quickly measure audience reaction, and get rid of the offender.

Chuck Wiese was not fired from KOIN-TV. He left for a full-time job (instead of part-time) with higher pay in Minneapolis. He had worked at KOIN-TV for 9 years. One does not survive that long on the air waves if the audience does not approve. Chuck moved to Seattle by invitation, again with a raise in salary. A news director at KING-TV had personal knowledge of Chuck's ability in forecasting, from the director's former experience in Portland.

If anyone has evidence of unprofessional or incompetent forecasting by Chuck Wiese, let me know. I am willing to amend my story.

"Dr. John" said:

"I have 34 years experience in weather forecasting as a qualified meteorologist.

"My doctorate is in Adult Education and included a dissertation dealing with the production of visuals for adults.

"Most people believe we actually make the forecasts. We don't. We work very closely with the NWS, and then report the information provided.

"Forecasters on TV should be in the primary job of *interpreting* as accurately as possible, using the most *attractive* visual aids as possible, in the most *understandable* way as possible, the NWS product."

7

The American
Meteorological Society

The AMS is tiny but important. The Society was founded in December 1919. Membership on December 31, 1946, totalled 2851. Membership in 1986 totals about 9800. The next largest meteorological organization is the National Weather Association (NWA) with a 1986 membership of about 1600. I have been a member of the American Meteorological Society since 1940 when I started as a Junior Observer for the Weather Bureau in Des Moines, Iowa.

The AMS headquarters in Boston, Massachusetts, is housed in an admirably restored structure called the Bullfinch House. Annual operating expenses of the Society are about $3,300,000.

The AMS is essentially a scientific society, and not a professional society in the ordinary sense of the word.

The constitution states:

The objectives of this Society are the **development and dissemination of knowledge** of the atmospheric and related oceanic and hydrologic sciences and the **advancement** of their professional applications. [bold print added]

Accordingly, the main activity is the publication of journals and books, and the conduct of scientific meetings and conferences. Such publications and books can be found in all university libraries, and many large libraries. The *Bulletin of the AMS (BAMS)* is the official organ. I have a file back to 1937.

Local Chapters of the AMS

The Society also grants charters to local Chapters, which

are autonomous affiliations of individuals who share a common interest with the national AMS. The Society provides guidance and assistance to Chapters, but "neither wishes to nor could exercise control or supervision over Chapter activities."

All officers of local Chapters, under the charter, are required to be members of the Society. Although Chapter members are encouraged to become members or associate members of the Society, such is not a requirement for Chapter members who are not officers.

The primary purpose of a Chapter is to serve the scientific and professional interests of its members in its own community. However, there are some mutual obligations of the Society and its Chapters to each other.

Responsibilities of the Chapter toward the Society include adherence to the policies, aims, and ideals of the Society. The Chapter and its members should always display the highest professional ethics.

The Oregon Chapter was started in 1947 by Dr. Fred Decker at Oregon State University at Corvallis. I helped Decker and others reorganize the Chapter and move it to Portland in 1955. In 1957 I felt honored to serve as president.

Boston Headquarters of AMS Viewed from Afar

I don't pretend to be knowledgeable about the national AMS, headquartered in Boston, but there are some disturbing signs viewed from afar. There is evidence of invasion by power-hungry bureaucrats. Science is not at the top of their priorities.

Since 1940 I have respected and admired the AMS, even to the point of hero-worship for some of its leaders. Rightfully, the AMS commands world-wide respect in scientific circles. Its prestige contradicts its small size.

Except for an executive secretary, a secretary-treasurer, and a small staff, the management of the Society is by unpaid elected or appointed members. The non-autocratic structure renders it vulnerable to invasion by any federal bureaucrats who might exist in the civilian National Weather Service and the military weather services.

Bear in mind that nearly all in the meteorological community are either on the federal payroll, or in academia, or in business catering to those groups. Of course, everyone on the federal payroll does not place agency self-interest, or military self-interest above all else. (This is my definition of a bureaucrat.) But a federal paycheck obviously influences one's perspective.

To understand the operations of the power structure, we not only must try to identify the individual bureaucrats, but we must also recognize their economic influence on non-bureaucratic colleagues. "Power goes to where the money is" is an axiom.

Academicians are teaching, or doing research, or managing. Most meteorology students are headed for federal careers. Most research is subsidized by federal grants. It is not prudent for academicians to criticize bureaucratic attitudes unless adequate political support can be found in non-bureaucratic circles.

Weather Tomorrow Society

Such political support will come soon from serious forecast users. SFUs are in a large blind spot of the weather world. Meteorologists have overlooked or forgotten about forecasting, which is the prime product of the technology.

I propose the creation of a **Weather Tomorrow Society (WTS)**, open to anyone interested in genuine weather forecasting (no other eligibility requirements).

This nation has at least 50,000,000 prospects for membership. Congress will listen to any large group which promotes constructive change.

With political support from the Weather Tomorrow Society, the American Meteorological Society can restore the authority and leadership of academicians, who ought to be, and used to be, in control of that society. The AMS can then continue to develop and disseminate knowledge, and advance professional applications.

Along a parallel path of progress the WTS can learn to recognize good forecasts and how to make use of them. The popular science interest in weather can prosper again (as opposed to popular pseudo-science).

But the main purpose of the WTS will be to keep the weather business under constant surveillance, and to serve the role of "enforcer." Bureaucratic minds in both the AMS and the WTS should be denied major leadership positions.

Revelations of Bureaucratic Invasion

The events of the last three years have opened my eyes. One thing leads to another.

To review, I protested the Mitchell Award. Dr. Lowry called back 46 days later. He had not yet forwarded the protest to Dr. Weinstein. Lowry declined to identify Phil Peck as nominator on grounds that information was confidential. Why should Peck's identity be confidential?

During the conversation Dr. Lowry complimented Dr. Weinstein, the new chairman of the Weather Forecasting and Analysis Committee, which selects the recipient of the Mitchell Award. Lowry told me, as though it were a credit, that Weinstein had been involved in the recent drafting of an AMS policy statement, which was "important." Yes, it was important to anyone who wanted to maintain federal control of forecasting.

I was yet unaware of such policy statement, so the comment didn't ring any bells.

After the anonymous threat to Chuck Wiese, and the historic meeting of February 6, 1984, I tried naively to alert Weinstein and the national AMS about the role of the NWS in the decay of local forecasting. Weinstein urged me not to look backward. I explained my intent was to funnel grass roots material to his committee, so it could guide the AMS Council.

The policy statement mentioned by Lowry was printed in the January 1984 *Bulletin of AMS*, under the title of "The Atmospheric Environment: an Agenda for Action." It strongly endorsed the domination of public weather forecasting by the federal government, which is the opposite of the intent of this book.

"A single source of official warnings of severe weather events must always be maintained." Weinstein helped draft that statement, which will be refuted in Chapter 9.

The Cylindrical Administrative Structure of Bureaucracy

Why should PhD researchers promote the federal monopoly on public forecasting? Because forecasting lends glamor to the weather business. Congress can more easily be persuaded to maintain the lucrative administrative structure of NWS and NOAA. One can guess that Lowry and Weinstein saw self-interest in that.

The usual metaphor of administrative structure compares it to a cone, with the number of executives decreasing at each higher level.

Dr. Fred Decker, with some experience in the Washington, D. C., scene, says government bureaucracies strive to create a cylindrical structure, with nearly as many Chiefs at each higher level as there are Indians below. The excess of Chiefs in the higher levels, with reduced workloads, fill their time communicating with each other.

The primary task is to keep Congress convinced this structure is necessary. In the weather business, the sacred cow is **protecting life and property**. These words stir emotions.

Observing weather conditions and assembling data do not in themselves relate vividly to protecting life and property. But **forecasting severe weather events** sounds almost identical. Therefore, the NWS strives to maintain an image of a savior against natural disaster to maintain its cylindrical administrative structure.

The Weather Tomorrow Society will help create a supply of genuine forecasts from private enterprise and strive to persuade Congress to trim the NWS cylinder to a cone.

There are silent members in the AMS who dare not risk a gadfly role. But those who are reasonably safe from possible retaliation should also join the WTS when it gets organized. Scientists, technicians, and SFUs can work side-by-side.

Complaints to AMS Charging
Unethical Conduct by John Walls

On January 30, 1984 John Walls told me:

> I've watched Chuck Wiese time, and time, and time again, walk into that TV station down there, and in 10 minutes has the whole thing figured out, and forecast something completely different from what the NWS was saying . . . That is really unethical, to say he can do that. I've been in this business for 30 years. I've never seen even some of the most outstanding forecasters in the country be able to do that.

I wonder how Walls observed Chuck so closely, when Chuck was working weekends, and Walls was working weekdays.

Walls again used the word **unethical** on February 6, 1984: [abbreviated]

Wiese   Your biggest objection was that I can come in, and in 10 minutes look at a weather map, and come up with a forecast completely different from the NWS?

Walls   Yes, I've seen you do that more than once.

Wiese   You also call this unethical?

Walls   Yes. I got it in writing. I've written it, and it's in your file. It's still there. It has nothing to do with the AMS.

These charges by John Walls were made before a group of weather forecasters, members of the AMS, in an office of the NWS, a public agency, with a tape recorder running. It was a controversial meeting, convened to "clear the air" after Chuck Wiese told me he received an anonymous threatening phone call.

Chuck stewed about this for awhile. In July 1984, he wrote to Walls demanding retraction of statements, and to desist from future harassment.   No reply.

On September 20, 1984, Wiese sent a formal protest to the national AMS at Boston about the unethical conduct of Walls. This was followed by several phone calls over a period of time.

I combed old issues of *Bulletin of AMS* for the history of the AMS Code of Ethics. As late as 1980 the By-Laws, Article VIII-B-2, stated: "He will not unfairly discredit his fellow professionals. However, he will present to the proper authority for action, information on unethical, illegal, or unfair practice on the part of members." That was deleted in late 1980.

However, the moral principle remains in Article VIII-A-1: "The member will conduct himself in a manner to reflect dignity and honor on his profession."

I also filed a formal complaint against John Walls on November 19, 1984, for unethical conduct toward Chuck Wiese. Maybe two complaints would get more attention than one.

The Boston headquarters never did give Chuck Wiese a formal response. The first response to me was dated February

25, 1986, 15 months after my complaint against John Walls. It was from Robert E. McLaughlin, legal counsel to the AMS, who reported that Dr. David F. Landigran, the Secretary-Treasurer of the AMS, had died unexpectedly on February 7.

On March 17, 1986, McLaughlin explained: "The AMS does not have the machinery, and can not hold itself out as a forum to settle all disputes among its members under the guise of enforcing a Code of Ethics. Your complaints and the complaints of Mr. Wiese may well be meritorious, but a professional and scientific association is not a court of law with the ability to properly adjudicate such disputes."

It is strange it took the AMS so long to so advise me. If the AMS can't express an opinion regarding an ethical problem as clear as this one involving Dr. John, then why does it maintain an Ethics Committee comprised of 5 members? Problems of ethics should not require adjudication. A society can enforce its code in any way it wishes.

The ethics of an organization are anything the organization wants them to be. The organization sets its own rules. Apparently the AMS has no standards, as such, but only a statement of ideals.

From my point of view, the leadership of the AMS was unwilling to admonish John Walls because he is a vigorous supporter of NWS policy to maintain its monopoly on public weather forecasting.

Therefore, I am taking this matter to the tribunal of public opinion. I don't mean the opinion within the tiny AMS, but the opinion of SFUs. Weather forecasting is a subject of national concern. I am convinced the meteorological community has become too insensitive to its supportive public.

The following explanation is hindered by lack of information from Boston. Dr. Landigran, the Secretary-Treasurer of the AMS, was a non-voting member of the AMS Council. He served as middle man between the Ethics Committee and the Council. He presumably forwarded any complaint to Richard G. Semonin, Chairman of the Ethics Committee, then relayed the committee's analysis to the Council for action.

Two letters from me to Semonin were not answered. Chuck Wiese made several phone calls to Landigran in Boston. Both Wiese and I talked with Semonin in Champaign, Illinois.

In April 1985 Landigran told Wiese "this is a difficult situation." In May 1985 he told Wiese he felt "pushed up against a wall," that although the AMS agreed some bad things had been done in regard to defaming Wiese's character and conduct, the AMS did not know what they could do about it.

On January 6, 1986, exasperated at long delay, I sent a letter individually to each of the 20 Officers and Councilors of the AMS, describing the stalemate. That apparently stimulated the decision relayed by McLaughlin.

I wrote to Dr. Kenneth C. Spengler, Executive Secretary of the AMS, "I am grieved and disturbed that Dr. Landigran passed away unexpectedly on February 7. He died one month after my letter to all Council members. I had gained, and retain, the impression he was an honorable man caught in the middle of something not of his making."

If the reader feels the above story a little unbelievable, he is invited to review the local story in Oregon.

My Complaint to the Oregon Chapter of the AMS

During 1985, George Miller, boss of the Portland WSFO, was president of the Chapter. At the last chapter meeting under his leadership he provided a 2-page handout (abridged below):

> Having served as president of two other local AMS chapters, I was looking forward to 1985, and serving as president of the Oregon chapter. The year proved rewarding and also interesting.
>
> During the year the Executive Council faced a different challenge. A local chapter member filed a formal complaint with the Council alleging unethical conduct on the part of two chapter members. [I had included James Wakefield in the complaint against John Walls]
>
> The Executive Council, acting as the Ethics Committee, solicited views from various individuals associated with the complaint. The Committee reviewed the evidence presented to it and dismissed the complaint. The responses were returned to the originators.

The management of the Oregon Chapter of the American Meteorological Society is placed in the hands of an Executive Council, elected each year in January to serve for one year. As far as I know, the above statement by Miller was the only information supplied by the Council to the Chapter members. Remember that my protest of the Mitchell Award in 1984 was hidden from the members of the Oregon Chapter.

The identity of the Council:

> George Miller, boss of WSFO, Portland
> Jim Little, weather reporter KGW-TV
> David Apple, weather reporter KATU-TV
> Phil Pasteris, hydrologist, NWS
> Jim Stanley, hydrologist, NWS
> Charlie Feris, forecaster, BPA
> Nancy Stephan, new meteorologist, BPA

Apple and Feris abstained from the vote to dismiss my complaint.

At a regular meeting of the Oregon Chapter on October 26, I handed out a 2-page report to members attending, and explained some of my views. Excerpts follow, which in turn include excerpts of the Council meeting on October 15.

That Council meeting was held during the lunch hour in the employees' lunchroom, at the Custom House, a public building. It was two weeks after the meeting on October 1, at which the Council dismissed my complaint. (Chuck Wiese was also present on October 15.)

> It is obvious this Chapter is unwilling or unable to correct wrongdoing. Errors are increasing rather than decreasing. Pursuit of short term goals of self interest will bring long term deterioration to the professional status of weather forecasting.
>
> This is a scientific society. Matters of professional ethics should be reviewed by any interested member. My complaint against Walls and Wakefield was open, but the rebuttals are kept secret.
>
> Excerpts of tape recording of Council Meeting on October 15:

Lynott   How can you reach a decision involving John Walls when presumably he didn't make any response, or any significant one. Did he appear before you?

Feris   No.

Lynott   Do you mean to say that you decided that he was right, and that I'm wrong, when you didn't even get any defense from him, for my accusations? That is the most absurd, bizarre thing I've ever heard of.

Miller   I can answer that, from my part. I have reams of information here, that has been sent out, with regard to John

Walls, and Bob Lynott, and Chuck Wiese. And I felt that from all that information, I could not see what else would add to that. In this particular information that I have, was in more ways than one, one-sided on your part. And I looked at that particular information in arriving at my particular decision. I did not just take--

Lynott   You didn't need any refutation from John Walls? No defense, you didn't need any defense?

Miller   In all of this material in here, there's plenty of defense from John Walls.

Wiese   From John Walls?

Miller   Uh huh. In all this written reams of material that I have.

Wiese   May we have copies of that? I don't think we've seen it.

Miller   No. In this crap Lynott has sent out!

Wiese   I don't know what crap you're talking about.

Feris   The difference transcripts of the different meetings, and the various material that Bob has sent us all, over a period of time.

Miller   That's the information I'm alluding to. I took that information into account in arriving at my decision.

Wiese   That information says things **against** John Walls. And then you decided **for** him, without any input from him?

Miller   They're copies of transcripts of recordings, of Bob's recordings, that say Chuck said this, and Bob said this, I said that. I don't consider that accurate.

Stanley   I see no problem with another formal complaint submitted to the committee. This time I would like to see a complaint specifically, and not a bunch of transcripts. I never did see **the** specific complaint written up.

Feris   There was a specific complaint.

Stanley   I've never seen it. That's all I'm saying.

Miller   Well, I think you did.

Stanley   There was so much garbage with it, it got lost in the shuffle. We were overwhelmed with paper.

Failure to respond to formal charges of wrongdoing, brought before appropriate authority, is called default. Stanley voted against my complaint even though he admits never reading it. Miller voted against my complaint without any input from Walls, and secret input from Wakefield. He claimed my evidence **against** Walls provided him with plenty of defense **for** Walls.

Presumably, the Council is anxious to keep the defenses of respondents secret to cover up the weakness of the defense. On October 1, Dave Apple said: "We have to make certain assumptions. No. 1. That all the people are honorable."

Earlier, Apple was scared that Walls or Wakefield might sue the Council.

Note the peculiar wording of the final motion by the Council: "Based on evidence provided to the Ethics Committee, it dismisses the accusations against Walls and Wakefield brought by Robert Lynott."

When something embarrasses you, and you can't escape it, just say "PRESTO" and it will disappear. You hope.

I accuse the Executive Council of acting as a Kangaroo Court.

Among the last words in George Miller's farewell handout, he said, "Meteorology is suffering from growing pains. It is our duty as meteorologists to provide others with knowledge about our science. A strong cohesive chapter of private and government meteorologists can do this."

If this were 1956 Miller might have been right about growing pains. In January 1986, at the meeting to elect new officers, I would diagnose the problem as arteriosclerosis in the bureaucracy of the NWS. Chloresterol is of national concern. Perhaps a reduction in the fatty fiscal diet of the NWS might help weather forecasting recover.

With political support from The Weather Tomorrow Society, the AMS can restore the authority and leadership of academicians, who ought to be, and used to be, in control of the AMS.

The NWS strives to maintain an image of a savior against natural disaster to maintain its cylindrical administrative structure.

"The AMS does not have the machinery to settle all disputes among its members under the guise of enforcing a Code of Ethics. A professional and scientific association is not a court of law with the ability to properly adjudicate such disputes." (AMS legal counsel)

"Meteorology is suffering from growing pains. It is our duty as meteorologists to provide others with knowledge about our science."   (George Miller)

8

Professionalism — What Goes On Here?

The ethics of an organization are whatever the organization wants them to be, and whatever it maintains them to be. Similarly, the professionalism of a federal agency is demonstrated by the behavior of its personnel, especially that of its local leaders.

In January 1985, at the annual meeting of the Oregon Chapter, when George Miller was elected president, I distributed a handout, a Proposal for the April meeting. Part is reproduced here:

> This chapter started in early 1947 at Oregon State College. Fred Decker was The Chairman through 1956. In 1955 he was the leader in a plan to redesignate the chapter to a statewide chapter.
>
> Decker listed seven "Useful Contributions" which could be made by the Oregon Chapter. The first two:
>
> 1. Drawing attention to the services which meteorology can give to various economic activities.
> 2. Emphasizing the ethical standards of the profession.
>
> The year 1955 was a time of beginning for meteorology here in Oregon. The situation was primitive and the spirit was robust. In the last 30 years, under the leadership of the AMS, the profession has made progress.
>
> Judging from the eagerness of certain people to use the title of meteorologist, we have indeed gained the respect of the public. We enjoy some measure of prestige.
>
> The prestige of our profession depends on the quality of our professionalism, that is, our methods, character, and standards. What will people think of us thirty years from now, in the year 2015 ? Will the label of meteorologist imply

respect, or will it be downgraded? What if it declines to approximately the same recognition that we now extend to the
word **bureaucrat**?

As topics for the next regular meeting I listed the recent
abuse of the AMS award system, the harassment of Chuck
Wiese, the imitation forecasting of the NWS, the apparent
domination of our Society and the academic community by
government bureaucracy, and the coming shakeup in
forecasting.

The newly elected Council, naturally, rejected the proposal as "inappropriate for an evening meeting. To discuss
such topics would obviously hurt attendance at future
meetings and dampen enthusiasm."

Meeting of Oregon Executive Council, April 16, 1985

However, the Council of the Oregon Chapter of AMS
did invite me to describe my concerns to them, at a meeting in
a public restaurant on April 16, 1985. It was the first of
several encounters. I had been promised one hour. To save
time I mailed six pages of fine print to each Council member.
This included the reopening of the Walls-Wiese adversarial
relationship. Here is an excerpt from a tape recording of the
Council meeting:

George Miller	Are members of the Executive Board unhappy with any of the ethics of our members?
Charlie Feris	Bob has some good points. The Walls-Wiese relationship —
George Miller	The Walls-Wiese relationship, I believe, is a matter between Chuck Wiese and John Walls.
Dave Apple	If we present to the membership all this information, how do we as officers determine it is presented in totally unbiased fashion? Every person has some axe to grind. Somewhere along the line, if you present this to the membership, you have to find blind justice. Do we have a blindfolded person who can say "Here are the unvarnished facts, no embellishments from any person's point of view."
	To present this to the membership

openly, it will almost require a large upscale grand jury type of investigation to determine what are the facts. Sure, Chuck got a threatening phone call. Who said so? Chuck Wiese. Was there a recording made of the conversation? No.

If you say to the AMS membership, one of your members has acted unethically, he has said thus and so. When you make an innuendo or accusation like that, you can become almost guilty of unethical behavior yourself if you besmirch someone else's reputation undeservedly, just by saying something that maybe was presented erroneously. So I think you are on thin ice.

My description of concerns presented to the Executive Council of the Oregon Chapter of the American Meteorological Society marked a significant enlargement of my gadfly role. It had begun with a protest of the Mitchell Award. Two successive presidents of the local AMS Chapter had declined to make a copy available to the members. The new Council was pondering what to do about my additional criticisms of local activities pertaining to weather forecasting.

George Miller reopened the controversy about the surprise snowstorm in Portland on December 20, 1984. He implied to the Council that I had accused his office of falsifying the records.

Falsifying records is rare or never in the Weather Service, but distortion in propaganda by certain local administrators is what Robin Cody obliquely described by "everything but oobleck for Portland."

I had visited the Weather Service Forecast Office (WSFO) on Monday, December 24, to check official forecasts. My letter-to-the-editor of *The Oregonian* criticizing double talk by the NWS was printed January 4, 1985.

George Miller came back with a contradiction in the same newspaper on January 14. He referred to a "Winter Storm Watch" that was issued the day before the snowfall. Such notice was not among the specific file of forecasts for Portland I had copied earlier.

On January 21, 1985, I again visited the WSFO to verify details of the Winter Storm Watch. It cost me $11.25 for "research" by Robert Roe to locate the item on the continuous roll of paper of the NOAA Weather Wire, by then in storage. It was another example of ambiguity. Misunderstanding between the NWS and me is now common.

There was an adversarial exchange between me and Phil Peck on that day of January 21, 1985, but Robert Roe exaggerated it. The roadblocks erected at the WSFO against my occasional search for data to confirm forecast errors discourages such effort, which apparently satisfies agency purpose.

Memorandum from Robert M. Roe

Three days after the Council meeting mentioned above, Robert M. Roe prepared the following memo. George Miller gave a copy to each Council member on May 7, 1985. Thus it was "published," revealing something about bureaucratic attitudes.

April 19, 1985 To George Miller From Robert M. Roe, Supervisory Meteorological Technician, Subject: Memo for the Record

On January 21, 1985, I was on duty in the public service section of the WSFO. It is the responsibility of this section to handle all manner of requests for climatological data, current and forecast weather information, and to research various types of products issued by this office that may be needed in support of investigations. There are, and have been, established procedures in regard to the assessment of charges for these types of activity and for the preparation of any copies of data that may be requested. The office copier is operated only by NWS employees.

I had occasion to overhear and observe Mr. Lynott and Mr. Peck in a very heated discussion in the foyer of the office. Without a doubt, Mr. Lynott was accusing Mr. Peck of covering up the lack of a winter storm watch issuance in December 1984, and of later fabricating one to take its place. I felt I could easily resolve the matter by checking the Weather Wire roll for the period in question.

Action: I suggested that I could retrieve the roll in question, search out the data, and answer Lynott's questions. As custodian of this roll, I did not feel that it was in the interest of Mr. Peck or the National Weather Service to allow Mr. Lynott to peruse this product on his own. I advised that I would do this for the applicable fee ($3.75 per quarter hour).

The Weather Wire roll, which is a few hundred feet long, was unrolled until the product in question was found and Mr. Lynott appeared to be satisfied. The roll was then rewound by hand by me. This is an extremely time consuming chore and I allowed Mr. Lynott to keep track of the time in regard to assessing the charges.

The actions that I took in retrieving the product in question for Mr. Lynott served to temporarily mollify him and relieved Mr. Peck of needless hassle that morning. Mr. Lynott appears to have great interest in the operation of this office but appears to be ignorant of how we operate. Comments in the letter provided to the MIC/AM by Mr. Lynott mention the fact that he has had previous visits to this facility in which he was allowed access to any forecast files that he might want. He also states that he made copies of "all forecasts pertaining to Portland for December 19, 1984, in the file." If so, I find no record of his having paid the fees normally charged for use of the NWS owned copier.

From conversations with other members of the WSFO staff, I have no doubt that Mr. Lynott was allowed such a free hand due solely to intimidation. In particular, I refer to an instance in the past year when Mr. Lynott and a Seattle T.V. personality came into the office demanding a look at certain files and running a tape recorder as they asked questions. The secretary, Mrs. Kellogg, was subjected to harassment and intimidation that is not required of any NWS employee. Demands were made for a roster of all station personnel under the "Freedom of Information Act."

While I have experienced it on milder scale than others, I can attest to the fact that Mr. Lynott has an extremely rude and abrasive manner when dealing with personnel from this office. It would appear that he feels he can subject a government employee to any type of harassment he wants to deal out with their having no recourse. I personally feel that he is abusing my rights under the Freedom of Information Act by demanding personal information without authorization.

I think Mr. Peck showed extreme restraint on January 21, 1985, in trying to deal in a rational manner with Mr. Lynott. As the manager on duty, Mr. Peck was correct in not subjecting Mrs. Kellogg to any more of Mr. Lynott's harassment and intimidation. I feel that Mr. Lynott is entitled to no more nor no less service than any other customer of this office. There are further channels he could pursue if we do not provide him with the answers to legitimate complaints.

It is strange Roe complained to Miller about my harassment of Mrs. Kellogg on April 17, 1984, because Miller was right there, and could easily have defended her.

I have a receipt for $4.50 for 15 copies which I made on December 24, 1984. (The price for photocopies at the WSFO is now 35c.)

The memo from Robert Roe did not come to my attention for several months. I demanded an apology from both Roe and Miller, which was ignored. I wrote to Hazen Bedke, the NWS Regional Director about the matter. He suggested I make an appointment with George Miller and identify the elements of my grievance. Miller had "served us well in the past, and I'm sure will serve you in the same manner."

Special Meeting June 20, 1985, to Discuss Professional Ethics

The Executive Council of the Oregon Chapter of AMS eventually arranged a special meeting on June 20, 1985 to give me an opportunity to voice my concerns about ethics and professionalism in weather forecasting. Meanwhile, I prepared a 36-page "Scrapbook" of material, and mailed it to all members of the Chapter. That would pave the way for group discussion, instead of a lecture from me.

Most of our Chapter meetings are semi-social dinner meetings. This one was different. Only 18 of 44 Chapter members attended, and 6 were Council members. There were three tape recorders, including mine. Dr. William P. Lowry, a Certified Consulting Meteorologist, and Secretary of our local Executive Council in 1983, was one of the first to enter the discussion [which is slightly edited].

Lowry

It seems to me I've heard at least two topics. One has to do with the mechanics of the AMS. The other has to do with the quality of forecasting,and there was this vague thing called bureaucracy involved in weather forecasting. The title of this seminar or whatever you want to call this, seems to me had to do with professionalism and forecasting.

So I for one would like a little guidance from whomever feels competent to give it, whether we're going to discuss both of those things or the second thing, and whether I'm incorrect in thinking that it's the second thing that we gathered here to discuss, since that is the title of the meeting.

So I leave you with a couple of questions. Anybody that

knows me knows those are not my last words this evening.

What I intend to do, whenever I speak, it's going to be concerning the quality of forecasting and the effect of bureaucracy on the quality of forecasting. That's what I came here to talk about.

If others want to talk on another subject maybe they will do as I am doing, and announce that that's what they came to talk about.

You made two statements, Bob, and I would like you to tell me what information you have, hard information, factual information, because I know it's available somewhere, and I presume you've looked into it.

Two statements you made, that in recent years the quality of forecasting has deteriorated. By what measure, and where did you gather that, ah, is that a gut feeling? Or is that documented?

The second thing you said is that bureaucracy and its stranglehold is the cause of that deterioration. I wonder if you have any basis for making that, other than a gut feeling?

So I'll start the discussion with two more questions. You know, as we have talked, on many things I agree with you. I'd like this thing to get down to a sharp point and I would like to see some kind of progress made if that's possible at this meeting tonight, rather than people just flapping their wings, and getting things off their chest.

Is that still a gut feeling you're expressing?

Lynott

You can hand out statistics until you're blue in the face, and still you have to make a subjective appraisal of the statistics.

Lowry

I'm playing Allan Murphy right now. I asked that question because he told me he would ask it if he were here. By what measure do you make that statement?

What is your basis for saying that bureaucracy is the cause of this thing? I think that you asserted that it is. What is your basis for saying —

Lynott

Bureaucracy is strongly involved in some of the acts of unprofessionalism in this area recently.

Lowry

Are we talking about ethics, or quality of forecasting? I thought we were talking about the quality of forecasting.

Lynott

We're talking about professionalism in its broad aspects, this includes competence, technical skill, and ethics, and other things. The forecasts made by the NWS on November 13, 1981, were poor.

The thing that I'm upset about, is that bureaucratically they wangled themselves a pat on the back by the AMS. That violates ethics.

Lowry

It's a giant leap from that to this graph that shows skill scores going down.

Lynott

Those are overall statistics that don't impress me much. Forecasting should be concerned with calling for significant changes which we don't get every day.

The meteorological community doesn't understand forecasting because it spins its wheels talking about average scores 365 days in the year, and they obscure their answers.

You should only look for the times there are significant changes and grade those. When you start doing that I'll pay attention to your verification scores.

Lowry

Tonight you are trying to make a case that it is important to them whether **you** pay attention to their verification scores? Why should they care what you think? Are you answering that question tonight? Is that why you're here?

I'll put myself in the case of this bureaucrat and I say, "He's one in 220 million people in this country. Why should I pay attention to whether he's gonna look at my verification scores? Who is this guy Lynott?"

Lynott

Why should anybody listen to Bob Lynott? Not for any position or authority that he holds because he has none. No position. No authority. I'm using logic.

And if my logic doesn't seem logical to you, then reject it. If it does seem logical to you, consider it. My only strength is logic.

It would seem there is a connection between the ethics of the Oregon meteorological community and the quality of forecasting. The local NWS is seeking to maintain its monopoly on public forecasting, which is in line with the policy of the national agency. The NWS is against the independent forecasting of Chuck Wiese. The thing called bureaucracy may be vague to Dr. Lowry, but it is not vague to anyone who wants to offer independent forecasts to the public.

Lowry emphasized "sharp points" and "hard information." From my point of view, this is the attitude of certain factions in academia toward subjective intellectual reasoning. Those factions are preoccupied with mathematical and statistical reasoning. I have respect for such reasoning, but it must be applied in a proper manner, not mechanically or automatically, and not by a robot.

Lowry said he was "playing Allan Murphy," who is a professor at Oregon State University, the alma mater of Lowry's PhD in meteorology.

Dr. Lowry seems to equate subjective reasoning with a "gut feeling." Subjective intellectual activity can hardly be "documented." But mechanical statistical tabulations are easily documented. Some are worthless.

The fragmentation in philosophy in the meteorological community is serious.

Here is another excerpt from the recording of the meeting on Professionalism.

Miller

We're somewhat violating rules by discussing something at which we don't have a motion for.

Earl Bates (retired agricultural meteorologist)

You can discuss any subject without a motion. When you have a motion you have to discuss it, or else vote on it.
[This awakened ideas to restrict discussion]

Jim Little (TV weather reporter)

I'm not sure where we're going tonight. Maybe if Bob Lynott would offer a motion, so we could focus the discussion, and ultimately get somewhere.

Lynott

Let's not bury ourselves in procedural details. Let's reason.

Little

Could we put this in the form of a motion, and talk about it?

Parliamentary antics gained control of the meeting. Jim Little moved that the Oregon Chapter look into my allegations about John Walls. Little rephrased his motion to form a committee to investigate the problem. Miller ruled the discussion must now be limited only to formation of a committee. The discussion continued: [slightly edited]

Apple

As the Treasurer, I think we should proceed with a committee very carefully. This is my fear. When you form a committee to look into the professional misconduct of a professional, and should that committee find reason to censure, or anything like that, and should it in some way affect the censured member's ability to do his job, or make a living in his profession, then there is the possibility that the committee may be found liable should an **error** be made through somehow missing a piece of vital information. Some of the officers of the AMS and some of the committee could be found liable.

You're again treading on very thin legal ice. Be careful! Because it's opening the door to suits. Use a little care in doing something that could destroy the Oregon Chapter with one swift stroke of the legal pen.

Be careful when you put a man's name on paper, and you're alleging professional misconduct, and you put it in the U. S. mail and mail it to members.

Lynott

Dave Apple is throwing out scare words.

Lowry

I drove all the way from McMinnville to talk about the quality of weather forecasts. If we're going to spend the rest of the evening talking about what has been named the Walls-Weezee, -Wiese affair, then I'm gonna leave.

Dave Apple is a weather reporter on KATU, Portland, a member of the newly designated public-private partnership defined by the National Weather Service.

The meeting degenerated into parliamentary confusion, defining a quorum, calling the Council into Executive session, and rewording the motion of Little, which eventually failed. At one stage Jim Little moved to define a quorum as **"members present."** Then he and Dave Apple changed it to **"majority of dues-paying members present"** (which didn't make sense either).

Discussion wandered into reminiscing about early experiences, and a discussion about competition in forecasting. Near the end of meeting, Jim Little spoke again:

Little

Who determines whether a forecast is useful or significant? As a television weathercaster I have a constituency of maybe 100,000 people every night. Each has something different that is significant to them. They have individual needs. There is no way I can address those individual needs.

And you are talking about significant changes. Sometimes significant changes are not what's important. If you're cutting alfalfa, and you don't want any change —

Mrs. Fred Decker:

But you want to know it isn't going to change.

Little

A large portion of folks have engineered weather out of their lives. They get in their car and they drive to a parking garage and the only time they're exposed to the weather is the walk from the parking garage to their office. How do you score the folks that are on TV?

Of course, a public weather forecaster can't meet the peculiar needs of each individual. But an accurate local forecast will satisfy 98% of serious forecast users in a specific area. The SFUs know what is significant.

It is strange to hear a TV weathercaster explain that many people have engineered weather out of their lives. Then why should they watch a weathercast? Possible answer: for the entertainment.

The NWS says it doesn't try to produce specialized forecasts for individuals or small groups. That proper admission should not be an excuse for ambiguous or inaccurate forecasts. SFUs must be alert for fuzzy semantics.

Dr. Allan H. Murphy, Professor of Meteorology, Oregon State University

Who is this professor from the Department of Atmospheric Sciences at Oregon State University? To begin, here is his letter to the Editor of *Northwest Magazine*, about the article by Robin Cody on July 28, 1985:

Your July 28 story, "Why Can't They Get It Right?" about the National Weather Service contains numerous unsupported or incorrect statements and reveals many misunderstandings. The accuracy of these statements as well as the nature of the misunderstandings could have been determined by contacting current staff members at Oregon State University--the site of Oregon's only academic department with a wide range of graduate programs in meteorology. Because of space limitations, a few examples of misstatements and misunderstandings must suffice here.

A basic thesis of the article evidently is that independent TV or private meteorologists can produce more accurate forecasts than National Weather Service meteorologists. This thesis may be true, but discerning readers will note that not one piece of scientific evidence is presented to support it. Comparison of the quality of forecasts produced by two groups is difficult because, unlike the weather service, private forecasters frequently do not verify their forecasts on a regular basis (or at least do not publish the results of their findings).

Mr. Lynott is quoted as saying, "Forecasting has not improved in 20 years." This statement simply is incorrect. Independent studies by me and other individuals reveal that weather service probability and temperature forecasts have improved significantly since the late 1960s. Other forecasts such as tornado watches also have improved in accuracy.

Mr. Anselmo's reaction to the probabilities used in National Weather Service precipitation forecasts is typical of that of many individuals who have not given careful consideration to the requirements of users who must make decisions--large or small--on the basis of forecasts. Such users need quantitative estimates of these probabilities to make the most efficient and effective decisions. Unless the certainty is quantified, the forecasts themselves become decisions--that is, the forecasters take on the added role of decision-makers, a role for which they generally are ill-prepared. Fortunately, experienced National Weather Service forecasters have demonstrated an ability to estimate these probabilities reliably and skillfully. Moreover, the public has indicated a strong preference for forecasts expressed in these terms.

Weather forecasting is undoubtedly one of the two or three most difficult and challenging scientific problems--a fact not adequately reflected in the article. Current efforts to make more specific forecasts on smaller space and time scales only increase these difficulties and challenges. To make the best possible forecasts and to communicate this information to the public and to specific users in a timely and effective manner requires, among other things, cooperation between National Weather Service meteorologists and their colleagues in the media and private sector--they all have important roles to play in this process.

Articles such as yours do not make a positive contribution either to cooperation and mutual respect among members of the meteorological community or to public understanding of the current status or future prospects of weather forecasting.

The academic credentials of Dr. Murphy are impressive. On pages 593 and 594 of the June 1980 *Bulletin of AMS* are 26.66 square inches of fine print describing his academic attainments. He was the 1980 recipient of the AMS Award for Outstanding Contribution to the Advance of Applied Meteorology. The award was given in recognition of "his innovative experimental and practical studies related to probability forecasting and for significant contributions to the theory and practice of forecast evaluation."

Dr. Murphy probably has published more pages in the publications of the AMS than anyone ever. I compiled a list of 50 references since 1965, and it probably is incomplete. One of them, in the *Journal of Applied Meteorology*, is entitled, **"On the nature of the non-existence of ordinal relationships between measures of the accuracy and values of probability forecasts: An example."**

A few comments about Murphy's letter-to-the-editor will suffice here. (Excerpts are in upper case; my comments are in lower case):

NOT ONE PIECE OF SCIENTIFIC EVIDENCE IS PRESENTED
    Failure on wind storm November 13, 1981; failure on snow December 20, 1984.
UNLESS THE CERTAINTY IS QUANTIFIED, FORECASTERS TAKE ON THE ADDED ROLE OF DECISION-MAKERS, A ROLE FOR WHICH THEY GENERALLY ARE ILL-PREPARED
    If the certainty is quantified (spelled out), is that not making a decision? What Murphy inadvertently revealed is that the robot in Washington makes the decision, not the local forecaster. Yes, NWS forecasters are ill-prepared to make

decisions about the weather tomorrow. Genuine forecasting is decision-making. Genuine forecasting includes a briefing, which explains the factor of certainty.

NWS FORECASTERS HAVE DEMONSTRATED AN ABILITY TO ESTIMATE THESE PROBABILITIES RELIABLY AND SKILLFULLY

They merely relay the estimates made by a robot in Washington D. C.

WEATHER FORECASTING IS UNDOUBTEDLY ONE OF THE TWO OR THREE MOST DIFFICULT AND CHALLENGING SCIENTIFIC PROBLEMS

How about cancer, alcohol and drug abuse, arthritis, nuclear hazards, and AIDS?

TO MAKE THE BEST POSSIBLE FORECASTS AND TO COMMUNICATE THIS INFORMATION IN A TIMELY AND EFFECTIVE MANNER

Anyone who uses the phrase, "timely and effective", has been studying the NWS propaganda manual.

COOPERATION BETWEEN NWS METEOROLOGISTS AND THEIR COLLEAGUES IN THE MEDIA AND THE PRIVATE SECTOR

Also a quote from the propaganda manual. (Of course, I am guessing. I have never seen the propaganda manual.) "Colleagues" are those who "cooperate" or acquiesce, either from self-interest or unwittingly, with the monopolisitic policies of the NWS.

ARTICLES SUCH AS YOURS DO NOT MAKE A POSITIVE CONTRIBUTION

Not for the bureaucratic cylindrical structure, that's for sure. Murphy has made a career of such positive contributions, and has gained recognition.

Dr. Murphy occasionally writes items for the OSU Department of Information. In one he said, "Weather forecasting is still an imperfect science, and the use of numbers occasionally may give an impression of precision in number predictions that simply does not exist." That's right, so a "discerning reader" is suspicious when Dr. Murphy later wrote:

Forecasts have economic value. Researchers at OSU are investigating just what that value is. Some verdicts are already in. For example, the current value of seasonal precipitation forecasts to spring wheat growers near Havre, Montana, is $4.06 per acre. If absolutely perfect forecasts of this type were available, they would be worth $79.14 per acre.

Gee, those researchers sure are scientific. They didn't say $80 an acre, they said $79.14. Is Murphy also an economist? If weather forecasts were perfect, crop yields would rise, but wheat prices then probably would fall.

Dr. Murphy also said:

> Many millions of dollars are spent each year in atmospheric research, and the federal government wants to know if the results are worthwhile.

The federal government is not really curious. But it is anxious that the taxpayers continue to provide funds for researchers like Dr. Murphy.

I happened to be in Corvallis one day when a special lecture was given by a researcher in the Department of Atmospheric Science. It was something about a computerized model for the atmosphere on Mars.

With a federal grant from the National Science Foundation, Dr. Murphy and Joel C. Curtis of the Seattle WSFO produced a 10-page scientific paper for the July 1985 *BAMS*: "Public Interpretation and Understanding of Forecast Terminology: Some Results of a Newspaper Survey In Seattle, Washington." The scientific analysis included the public interpretation of phrases like "Mostly sunny." Important topics are worthy of government supported research.

Additional Response to "Why Can't They Get It Right?"

Professional George Miller was the first to get his letter-to-the-editor printed after the appearance of Robin Cody's article on July 28, 1985:

> Your July 28 story, "Why Can't They Get It Right?," by Robin Cody, was particularly demeaning to the National Weather Service as well as to weather forecasting and reporting in general. The cover to the magazine was done in poor taste. Even if the question "Why Can't They Get It Right?" were true, the article did not address that issue at all.
> Cody's inference that the National Weather Service and television meteorologists are in collusion is totally inaccurate. We do meet at least twice a year, but to discuss advances in the science, new products, release times, etc.
> In Cody's interview, I presented factual statistics as to how forecasts are verified nationally and how we verify in-

dividual forecasters (National Weather Service) locally. I specifically pointed out how we improve upon our guidance. The article mentioned none of this, but suggested we forecast everything so as not to be wrong. Absurd.

I am the first to agree that there is a definite place for private meteorologists in the weather business. I encourage their endeavors. Yet to infer that they are always right and that the National Weather Service is always wrong is preposterous.

My "conservative" nature prevented me from acting irrationally and making wild assertions. Perhaps others need to do this and leave weather forecasting and reporting to the professionals.

Notice that Miller twice linked the phrases "weather forecasting and reporting," at the beginning and end of his letter. The NWS strives to blur the distinction, by implying they are closely related. Actually, they are far apart.

The article "Why Can't They Get It Right?" addressed the issue of weather forecasting head on. The NWS has been conducting a propaganda campaign for many years, with no opposition. It was so easy. The machinery was exclusively in their hands. Counter-propaganda from Robin Cody is viewed as "irrational and making wild assertions."

Concerning the "collusion" (which may have been overstated, depending on the definition of the word) between the NWS and media weathercasters, why are some of the latter opposed to an effort to revive the art of forecasting? At present, nearly all of them are getting their forecasts from the NWS. If a non-federal forecaster can offer more accurate forecasts (which must be demonstrated, of course), for dissemination by the weathercaster, why shouldn't the media leap at the opportunity?

Miller claims his staff improves upon guidance. Can he show one example of Portland forecasters significantly improving on the robot guidance?

The "definite place for private meteorologists" means a place that does not intrude into the monopoly of the NWS on public forecasting. The media weathercasters are not intruding. They are disseminating the federal product, even though usually pretending it is their product. Any subservient weather reporter thus can be elevated to a "private meteorologist.'' Hence such phrase loses distinction.

Phil Peck, Supervisory Forecaster at the Portland WSFO, and the nominator for the Mitchell Award for the 1981 windstorm forecasts, also wrote to the editor:

> Your story, "Why Can't They Get It Right?" deserves an extensive reply in order to correct all the malicious innuendoes and basic misinformation contained in it. For the sake of brevity, I will point out just a few of the inaccuracies that typify the article, rather than refute, point by point, all of the erroneous and unfair statements contained in Robin Cody's one-sided story.
>
> The article begins by belittling the warnings issued by the National Weather Service for the windstorm on November 13, 1981. The fact is that warnings were issued well in advance of the high winds and we--the National Weather Service--received many compliments from individuals and organizations who took protective action. For example, Flightcraft at Portland International Airport was able, between 6 p.m. and midnight, to tie down about 100 aircraft and thereby greatly minimized potential damage, because of our early warning. Two weather service forecasters received the Charles D. Mitchell award for exceptional forecasting of this storm, an award given annually by the American Meteorological Society headquartered in Boston. Bob Lynott disagreed with this award, which is his right, and he took his case to the society, which is the organization in charge of professional ethics in meteorology. It reviewed the evidence and upheld the award. This of course, was not mentioned by Cody.
>
> Cody's article has a private meteorologist implying that the use of probabilities in weather forecasting is absurd. This charge is simply ridiculous. Probabilities are the language of uncertainty. The probability of an impossible event is zero, and that of a certain event is 100 percent. Predicting future events precisely seldom is done with absolute certainty. So probabilities express the degree of certainty (uncertainty) of a future event. Probability numbers are the most precise means of communicating what the forecaster has in mind.
>
> On the matter of incentive: the cirriculum leading to a degree, required to become a professional weather forecaster, consists basically of math and physics. One enters this field almost always because of unusual interest. It is not a course taken because the student doesn't know what else to take or because it's an easy degree. This

unusual interest *does not* change or diminish depending on the employer, and it is patently false to claim that one form of employment promotes superior forecasting over another form of employment.

The article characterized Lynott's work as a weather forecaster as "often mistaken but never in doubt." On this I do not quarrel. In fact, this also is a fair description of a kibitzer, which Lynott well could be considered.

Were Cody truly interested in assessing the current state of weather forecasting, he might have thought of visiting Oregon State University. Its Department of Atmospheric Sciences is recognized as one of the finest in the country and could be counted on to give some interesting, if not downright authoritative comments and answers. Cody did not bother to do this.

"Why Can't They Get It Right?" is a question similar to "When Did You Stop Beating Your Wife?" The inherent bias of the statement gives evidence that the questioner is not interested in facts, just accusations. If The Oregonian wants to support and to be the medium for such slanted journalism. I'd rather not read it or pay for it. Cancel my subscription.

Peck claims the national AMS is "in charge of professional ethics." The AMS told me, "The AMS does not have the machinery, and can not hold itself out as a forum to settle all disputes among its members under the guise of enforcing a Code of Ethics."

In reference to the Mitchell Award, Peck claims the AMS "reviewed the evidence and **upheld the award**. This was not mentioned by Cody."

I will mention it again here. Dr. Dale Lowry (an earlier friend of Peck) called me, and expressed dismay at the evidence which I had forwarded (31 pages).

Dr. Weinstein wrote: "I won't debate the wisdom of the award . . . What is done is done! There is nothing to be gained in further airing of the past."

The AMS didn't "uphold" the award. It simply didn't rescind it, which no one should expect, and which I didn't expect.

Jim Little, weather reporter at KGW-TV, had a different concern than Peck:

Thanks for the article on weather forecasting. The role of private vs. governmental weather forecasting has been the subject of discussion in the meteorological community for many years. It's good to see it discussed in a more general forum.

I was interested to note that while Cody made some rather sweeping generalizations about television broadcasters, he made no attempt to interview someone he described as one of Portland's "heavy hitters"--me.

Echo of an Old Refrain, "Chuck Wiese, Troublemaker"

The NWS delivers its "products" over a statewide teletype circuit called the NOAA Weather Wire. Chuck Wiese appears on KING-TV in Seattle at 5 p.m. and 11 p.m. on Saturday and Sunday. No other weather program appears as early as 5 p.m. on those days. He also prepares forecasts for KING radio and TV on Thursday and Friday of each week.

Several times when Wiese has scooped the NWS, some disgruntled NWS staff member has delayed the delivery of temperature tables for Washington beyond the scheduled time of about 4:30 p.m., too late for use by Wiese to prepare graphics for his 5 p.m. show. Wiese has repeatedly complained to the WSFO, which denies any delay.

In midafternoon on Saturday, November 16, 1985, Wiese aired a forecast over KING radio for snow sticking on the ground in Seattle area, contradicting the forecast from NWS. He did not receive the temperature tables in time for his 5 p.m. TV show, and explained that fact to the TV viewers.

Apparently the Seattle WSFO monitors Chuck's shows. After 5:30 p.m., Robert Jackson, forecaster, called Sharon Callis, an assistant in KING newsroom, claiming the tables had been sent on time.

Transmissions on the NOAA weather wire can be controlled. A single address can be omitted from a transmission to all other subscribers.

Jackson again called KING on Monday, an off-duty day for Wiese. Jackson reached Bill Berg, assignment editor. Jackson claimed Wiese had made complaint over the air because Wiese had once applied for a job with the NWS and was rejected for employment, and was trying to cause trouble.

Wiese demanded a written apology from Robert Jackson, because Wiese had never applied for employment with his long-time adversary.

News directors don't understand the things revealed in this book. But they are susceptible to rumors. That's why, when Wiese first started working in Seattle, an anonymous caller warned KING-TV that Chuck Wiese was a troublemaker.

This is the letter Robert Jackson sent to Chuck Wiese: (slightly condensed)

> On Saturday evening, Nov.16, I telephoned KING-TV after viewing your weathercast, and made some statements which I have since regretted. My comments resulted from frustration concerning your on-the-air statement that you could not give the "max temperatures for today because the Weather Service did not send them to me." This was not true, and our records indicate they were sent at 4:21 p.m. PST.
>
> I realized as soon as I hung up the phone that I should not have taken your comments so personally. In the future, comments I have concerning your broadcast will be channelled through the Meteorologist-in Charge. Please accept my apology, and I regret any embarrassment my comments may have caused you.

Wiese had suffered no embarrassment. Also, he has reported no similar incident since November 1985.

If Jackson realized his over-reaction after the first call, why did he call a second time two days later?

"Two or More Councilors"

Small events sometimes illuminate attitudes.

The national AMS wisely declares it "neither wishes to nor could exercise control or supervision over Chapter activities." The national AMS can establish objectives and provide guidance, but local activities are the responsibility of the local meteorologists (local people in the weather business).

I helped organize the Oregon Chapter of the AMS in 1955. In my opinion, during the years, it has increasingly become dominated by the bureaucrats in the National Weather Service. However, in my opinion, a considerable segment of forecasting personnel at the Portland WSFO, and probably all hydrologists in the Portland River Forecast Center, place AMS objectives above agency interests. Hence they are not bureaucrats. But it is the undue influence of a few bureaucrats that must be revealed.

At the first meeting of the newly elected Executive Council on January 30, 1985, George Miller, president, (according to the minutes) "stated the need to amend . . . the Chapter constitution to include **at least two** councilors."

The constitution provides for a local Council of 7, which seems enough for a Chapter of 40 or so. The Council is composed of last year's president, 4 officers with titles, and 2 without titles. National rules require all officers to be members of the national AMS. The Oregon constitution is similar, but contains an obsolete reference to "professional members."

Presumably, Miller wanted more people on the Council, but not necessarily national AMS members. Subsequent events are partly hazy. John A. Jannuzzi, forecaster at the Portland WSFO, was temporarily on the Council before transfer to Seattle, and he was involved in wording the proposed change.

The final proposal was for **two or more** Councilors, and the President, the Vice-President, the Secretary, and at least one Councilor shall be professional members of the AMS (a category abandoned some years ago), and the other Councilor(s) and the Treasurer **should** hold some grade of membership in the AMS.

The Council could now be of any size, even the entire Chapter membership, and only 5 would need belong to the national AMS.

Amendments to the constitution must be by mail ballot. At the regular meeting on June 1, and in front of our distinguished guest speaker, Dr. John V. Byrne, President of Oregon State University, and recent head of the National Oceanic and Atmospheric Administration, George Miller passed the motion by a show of hands.

At the special meeting on professionalism on June 20, Miller announced "It appears as though we didn't have a quorum on that particular night, so anything we did was null and void with regard to changing our councilors from two to three. So we still have two councilors."

The Chapter has no quorum for meetings because business affairs are conducted by the Council. Also, the proposed amendment did not contain the word "three."

I wrote to Jannuzzi in Seattle, trying to unravel the

history of the event. Jannuzzi replied:

> There were two main reasons for seeking change in the local By-Laws. One was an attempt to diversify the council thus bringing in new insight. The second one was not to exclude so many local chapter members from involvement. The last nominating committee had a difficult time getting candidates to agree to run for office. They found that many were not members of the national AMS. I felt that with the four primary officers all being National AMS members it might balance the group to have at least some of the councilors non-national members.
>
> I really can't give you answers of the detail you are seeking. As to whether I or someone also used the word "should" or "shall," I haven't the faintest notion (nor do I understand why anyone but a Philadelphia Lawyer would care.)
>
> Now I would like to get something off my chest. You seem to spend all of your effort trying to find fault with events that have already taken place. I would counsel you, as my father did me, and I am to my children: to direct your efforts toward constructive, not destructive ends. The world is full of people trying to tear things down. There is a shortage and dire need of talented individuals to build things up.

The professionalism of a federal agency is demonstrated by the behavior of its personnel, especially that of its local leaders.

The roadblocks erected at the WSFO against my occasional search for data to confirm errors discourages such effort, which apparently satisfies agency purpose.

"My conservative nature prevented me from acting irrationally and making wild assertions. Perhaps others need to do this and leave weather forecasting to the professionals." (George Miller, responding to Cody's article)

"If *The Oregonian* wants to support and to be the medium for such slanted journalism, I'd rather not read it or pay for it. Cancel my subscription." (Phil Peck, supervisory forecaster, Portland WSFO)

9

The
Public-Private Partnership

This chapter illustrates dissembling and distortion by the weather bureaucracy. The intended meaning of the "public-private partnership" is the presumed cooperation between federal meteorologists and non-federal meteorologists.

The quotations may be tedious. The reader may skip this chapter if already convinced that the so-called partnership is propaganda from the NWS.

Harmony within a profession usually is a sign of stability and maturity. Responsible members usually strive for harmony. However, disagreement is legitimate if it is open, and if proper steps are taken to resolve any problem. This book brings problems into the open and urges resolution.

Partnership usually implies equality and freedom of choice. In this case the power is lopsided by a ratio of 1000 to 1. Freedom of choice never existed. The public-private partnership is a myth, asserted and enforced by an arrogant bureaucracy, and meekly accepted by most non-federal meteorologists.

The discussion here relates to public forecasting. The NWS claims 100% exclusivity in public forecasting, although Congress never granted exclusivity by law. Because of long-term propaganda, most citizens have been brainwashed to accept such belief without questioning. Even the leaders of a scientific society are misled.

"AN AGENDA FOR ACTION"

On September 29, 1983, the American Meteorological Society issued a policy statement, **"The Atmospheric Environment: An Agenda for Action."** It was partly authored by Dr. Alan I. Weinstein.

The policy statement required five large pages of small print in the *Bulletin of AMS* for January 1984. It reeks of federal bureaucratic propaganda to maintain and enlarge the cylindrical administrative structure.

The statement deliberately blurs the fact that the private sector now has two distinct segments — those forecasters serving private clients for a fee, and the newly designated media weathercasters. Careful reading will reveal a subtle switching in meaning as needed to serve the propaganda.

(Excerpts, sometimes condensed, are shown in upper case; my comments are in lower case. Bold print added.)

RECENT ADVANCES HAVE BROUGHT THE ATMOSPHERIC SCIENCES TO THE THRESHOLD OF AN ERA. DRAMATIC IMPROVEMENTS ARE WITHIN REACH IN THE PRECISION OF WEATHER FORECASTS. THE AMS HAS DRAWN UP AN AGENDA FOR ACTION TO INSURE THAT IMPROVEMENTS KEEP PACE WITH THE GROWING DEMANDS OF SOCIETY.

THE PROTECTION OF LIFE AND PROPERTY AGAINST THE HAZARDS OF NATURE IS AN OBLIGATION OF THE GOVERNMENT. ACCURATE WEATHER FORECASTS FOR THE PREPARATION OF COMMUNITIES TO CONFRONT SUCH HAZARDS IS ESSENTIAL.

The meteorological cancer described by Snellman has brought us to the **end** of an era, not the threshold.

MILLIONS OF INDIVIDUALS MAKE WEATHER-BASED DECISIONS EACH DAY, MORE EFFICIENT OPERATIONS BY THE GOVERNMENT AND THE PRIVATE SECTOR SHOULD BE A NATIONAL GOAL.

The linking of "government and the private sector" suggests a partnership. The linking with "millions of individuals" implies that the new group of media weathercasters is considered part of the private sector. The weathercasters obviously have access to the millions of individuals.

However, independent forecasters (the original private sector) are effectively denied access to the millions of individuals (which translates to millions of dollars).

Of course, if they can find a few individuals willing to pay for a non-public forecast, using a door-to-door selling effort, that is okay. The independent forecaster must show

that the Avon® lady has a better product. But her budget
for advertising is much smaller than the NWS budget for
propaganda supplied by the taxpayers.

A STRONG INSTITUTIONAL FRAMEWORK IN THE FEDERAL
GOVERNMENT IS REQUIRED TO FACILITATE
POLICY FORMULATION, SCIENTIFIC RESEARCH,
TECHNOLOGICAL DEVELOPMENT, AND THE PROVISION OF
WEATHER SERVICES.

What is meant by weather services? A good case can be
made for data acquisition and computer products by the
government.
    But "forecasting" has also been highlighted. If the
AMS also means forecasting services, then why not also
strive for a socialistic framework to take over operation of
transportation systems, farming, the stock market, and
McDonalds®?

PRIMARY RESPONSIBILITY FOR GENERAL WEATHER INFOR-
MATION RESIDES WITH GOVERNMENT.

Okay, if "information" does not include weather forecasts.
But the bureaucracy continually infers that predictions also
come under the category of information.

THERE IS AN IMPORTANT AND GROWING ROLE FOR THE
PRIVATE SECTOR AND A SUSTAINED PARTNERSHIP OF THE
PUBLIC AND A STRENGTHENED PRIVATE SECTOR WILL BE
NEEDED.

Based on the recent experience described in this book,
they don't mean private-sector Chuck Wiese, they mean
private-sector John Walls.

WE ARE WITNESSING AN EXPLOSIVE GROWTH IN
TECHNOLOGY, AND INFORMATION DISSEMINATION. VAST
AMOUNTS OF DATA ARE REQUIRED FOR QUICK-RESPONSE
FORECASTS AND WARNINGS. WE CAN DISSEMINATE THE
RESULTS TO THE PUBLIC RAPIDLY AND EFFECTIVELY.
TELEVISION DISPLAYS IN THE HOMES OF MILLIONS ARE AN
EVERYDAY DEMONSTRATION OF THIS TECHNOLOGY. LARGE
COMPUTERS PRODUCE WEATHER FORECASTS.

Presumably, Dr. Weinstein dug out "The Unchained Goddess" and reviewed it. Note the propaganda trademark of
combining "forecasts and warnings," and this time "quick
response" was added. The NWS recognizes the efficiency
of TV for public dissemination, and has proceeded to

dominate weather shows. Lastly, it unabashedly admits its forecasts are made by a robot.

PROTECTING LIFE AND PROPERTY: SIGNIFICANT IMPROVEMENTS IN THIS PRINCIPAL MISSION ARE NOW POSSIBLE. WARNINGS OF SEVERE WEATHER PREVENT LOSS OF LIFE AND PROPERTY. WE NOW KNOW HOW TO DESIGN AND CARRY OUT RESEARCH EXPERIMENTS THAT CAN IMPROVE UNDERSTANDING AND FURTHER IMPROVE WEATHER WARNINGS. WE CAN NOW PINPOINT DESTRUCTIVE WEATHER PHENOMENA IN SPACE AND TIME MORE ACCURATELY.

ENOUGH TECHNOLOGY IS AVAILABLE TO COMMENCE PROCUREMENT OF SYSTEMS TO PROVIDE IMMEDIATE IMPROVEMENTS IN WARNING SERVICES. A PROGRAM OF RESEARCH WOULD IMPROVE OUR PREDICTIVE CAPABILITY. INTENSIVE FIELD PROGRAMS ARE DESIRABLE TO OBTAIN NECESSARY DATA FOR THIS RESEARCH. ACCESS TO GREAT COMPUTING POWER WILL BE REQUIRED TO PROCESS DATA, TO TEST THEORETICAL CONCEPTS, AND TO DEVELOP FORECAST PROCEDURES.

This persuasive propaganda reveals the mind-set that forecasting is dependent on more data, more systems, more computers, and more PhDs to manage them. This inexpensive book offers a time-proven concept — that a few human minds with ability to interpret weather maps, and make subjective conclusions, will usually scoop the robot, and cost less money.

The wise heads in the AMS need to shake the bureaucratic monkeys off the back of that scientific organization, and steer it back into an intellectual course.

SERVING THE ECONOMY: IMPROVEMENTS IN BOTH THE GENERAL AND SPECIALIZED WEATHER SERVICES PROVIDED BY THE NWS CAN BRING IMPORTANT ECONOMIC BENEFITS TO THE NATION.

This inadvertent slip with the word "specialized" refutes the frequent claim that the NWS does not try to compete unfairly with the private sector, here meaning independent forecasters.

THE TECHNOLOGY IS AT HAND WHICH PERMITS ACCESS TO ALL WEATHER INFORMATION BY PRIVATE METEOROLOGICAL COMPANIES. THE FEDERAL GOVERNMENT IS URGED TO PROVIDE THE ARRANGEMENTS TO FACILITATE ACCESS TO THE OBSERVATIONS AND COMMUNICATIONS NECESSARY TO EN-

COURAGE THE GROWTH OF PRIVATE METEOROLOGICAL SERVICES.

This is parallel to providing access for the trucking industry to the national highway system. Unstated is the autocratic determination by the government that independent **forecasting** services will not have access to the mass market, the exclusive sales territory of the bureaucracy, that gold mine which provides justification for its cylindrical administrative structure.

WE MUST MAINTAIN AND STRENGTHEN THE KNOWLEDGE BASE. UNIVERSITY TEACHING PROGRAMS NEED TO DEVELOP SCIENTISTS TRAINED IN THE USE OF ADVANCED MEASUREMENT AND COMPUTER TECHNOLOGIES.

Which they presently do. But no mention is made of training in subjective decision-making, which is woefully lacking in the meteorological community, proof of which is the scarcity of forecasting talent.

THE ATMOSPHERIC EDUCATIONAL ENTERPRISE IS NOW ENCOUNTERING SIGNIFICANT PROBLEMS BROUGHT ABOUT BY THE TECHNOLOGICAL REVOLUTION. UNIVERSITIES NEED THE HUMAN-COMPUTER INTERACTIVE SYSTEMS TO ASSIMILATE, MAP, AND DISPLAY SIMULTANEOUSLY MANY TYPES OF DATA, WITH REAL-TIME CAPABILITY.

Whoever wrote that last sentence was a self-dazzled showoff. But it does reveal why we have technological problems. The so-called forecasters are bewildered with too many Tinker Toy® machines. Weather forecasts are made by logical minds, not by machines.

THE PUBLIC-PRIVATE PARTNERSHIP: WEATHER SERVICES SHOULD **CONTINUE** TO BE PROVIDED BY A **MIX** OF PUBLIC AND PRIVATE SECTOR SERVICES.

Clever deception: 999 parts public, 1 part private is not a "mix." One cannot continue what never existed.

GENERAL WEATHER SERVICES ARE PRINCIPALLY THE PROVINCE OF GOVERNMENT, AND SERVICES FOR INDIVIDUAL USERS ARE PRINCIPALLY THE PROVINCE OF THE PRIVATE SECTOR. HOWEVER, **A SINGLE SOURCE OF "OFFICIAL" WARNINGS OF SEVERE WEATHER EVENTS MUST ALWAYS BE MAINTAINED.** THE PUBLIC/PRIVATE PARTNERSHIP THAT HAS DEVELOPED OVER THE YEARS SERVES THE NATION WELL.

This is the core of the NWS-AMS self-serving propaganda. The Public-Private Partnership is a myth. By endless repetition this propaganda theme has become dogma, with an ambition to become tradition.

THE PRIVATE SECTOR NOW SERVES SPECIFIC USERS. THIS FUNCTION SHOULD CONTINUE TO BE RESERVED TO THE PRIVATE SECTOR. THE GOVERNMENT MUST ENCOURAGE THE FURTHER DEVELOPMENT OF SUCH PRIVATE SECTORS.

This has a fair-minded ring. But the part reserved is nearly infinitesimal, and shrinking rather than developing.

THE GOVERNMENT WEATHER SERVICES SHOULD SERVE THE GENERAL PUBLIC AND BROAD SECTORS OF THE ECONOMY SUCH AS AVIATION, AGRICULTURAL, AND MARINE INTERESTS, WHERE SUCH SERVICES ARE NOT TAILORED TO OR INTENDED FOR INDIVIDUAL USERS.

The "weather services" they are talking about also include forecasting, not just data collection and weather reporting. The only answer to this assertion is **"why?"**

IN THE SEVERAL FUNCTIONS OF OBSERVING, COMMUNICATION, PROCESSING, FORECASTING, AND DISSEMINATION WHICH MUST BE PERFORMED TO RENDER SERVICES TO USERS, THE GOVERNMENT AND THE PRIVATE SECTOR HAVE NATURAL AND LOGICAL DIVISIONS OF RESPONSIBILITY.

The division proposed in this book is much more logical. Forecasting requires only about 1000 special-type minds to serve the entire nation. If dissemination by the private sector via radio and TV is logical, why should not independent forecasting via radio and TV seem even more logical?

ONLY THE GOVERNMENT CAN MAINTAIN THE INTEGRITY OF THE OBSERVATION, COMMUNICATION, AND DATA PROCESSING FUNCTIONS.

True, as it does for defense, fire, police, public schools, highways, and the judiciary. But obviously it has not maintained the integrity of forecasting. Failure is glaring.

THE PRIVATE SECTOR HAS THE PRINCIPAL MEANS AT ITS DISPOSAL TO DISSEMINATE GENERAL FORECASTS AND WARNINGS THROUGH MEDIA SUCH AS RADIO AND TELEVISION. IT DISCHARGES THIS FUNCTION WELL AND ON A SELF-SUSTAINING BASIS.

On the previous page the AMS asserted that "the private sector now serves specific users." It switches here to include the media weathercasters as part of the private sector.

THE RELATIVE RESPONSIBILITIES OF THE PUBLIC AND PRIVATE SECTORS FOR SEVERE WEATHER WARNINGS REQUIRE RESOLUTION BY ALL PARTIES IF THE PUBLIC IS NOT TO BE CONFUSED IN POTENTIALLY DISASTROUS SITUATIONS. THERE NEEDS TO BE AN "OFFICIAL" WARNING RECOGNIZED AND DISSEMINATED BY ALL. THIS OFFICIAL WARNING IS THE RESPONSIBILITY OF THE NWS. THE PRIVATE SECTOR HAS AN OBLIGATION TO CONVEY SUCH WARNINGS PROMPTLY WHEN ISSUED. **THE AMS CALLS ON THE FEDERAL GOVERNMENT TO TAKE THE INITIATIVE IN DEVELOPING AN EFFECTIVE RESOLUTION OF THIS INCREASING PROBLEM.**

This is a desperate appeal by the NWS and AMS to persuade the federal government to pass laws giving the NWS exclusive rights for public forecasting.

This is convincing evidence of the invasion of a scientific society by self-serving bureaucrats. Every scientific-minded member of the AMS should feel ashamed.

THE AMBITIOUS AGENDA OUTLINED HERE REQUIRES A PARTNERSHIP OF EFFORT, AND WISE POLICY DECISIONS BY THE CONGRESS AND THE EXECUTIVE BRANCH.

Serious forecast users should pray that Congress will be wiser than the bureaucrats in the NWS and the AMS in making certain decisions. SFUs should make sure that all senators and congressmen become aware of the arguments in this book.

This nation does not need a big new batch of expensive equipment and bureaucratic administrators.. It does need to encourage a small batch of Alan Kent second-rate minds to produce genuine forecasts under the banner of "private enterprise", and without any harassment from bureaucrats.

SFUs succeeded grandly in upgrading forecasting after World War II. They can succeed even more grandly in the next few years because electronic technology, and our knowledge of atmospheric physics, are more advanced. SFUs must restore "man" to the "man-machine mix."

This nation did not explore the surface of the moon by using robots. We ought not rely entirely on them to make our local weather forecasts.

The Myth of "Public Confusion"

Of course, public confusion about weather forecasts must be avoided. The only possible confusion would occur if the public is given contradictory forecasts when severe weather is impending, or even contradictory forecasts when severe weather is not impending (for example, a false alarm).

Furthermore, the public will not be misled or confused by a "contradictory" forecast issued from a source without a reputation for reliability. Hardly anyone listens to those pitiful figures who proclaim the world is coming to an end tomorrow.

How best can such confusion be avoided? The federal bureaucracy claims the public should only receive a forecast from the NWS. That would be convincing if there was confidence that such forecasting was the best possible. But such confidence is not justified, as this book tediously explains.

Where else might a forecast originate? That is a temporary weak point in my argument. Temporarily there is a shortage of skilled forecasters. But this can soon be rectified if harassment ceases on the part of the NWS.

Right now, the job is to educate the public about the deplorable state of affairs, and to rally political support for constructive changes.

The job of a referee is time-consuming. After the publication of this book I intend to monitor the forecast products of the NWS in this area, and prove the incompetence of NWS local forecasters. I also will try to persuade TV stations to subscribe to genuine forecasting services.

Eventually there will be no public confidence in NWS forecasts, and hence no public confusion. I don't know the details of the situation in Minneapolis, but I assume the WSFO of the NWS does not depart significantly from the predictions of WCCO and KSTP. They follow rather than lead. Then why should taxpayers pay for forecasters that follow?

There doesn't seem to be any public confusion in Minneapolis. Also, I would guess that in such a weather-eventful place, 60% of the population are SFUs.

In fairness, let us assume that WCCO and KSTP often issued incorrect forecasts of severe weather, or even forecasts for significant changes not really impending. They would quickly lose credibility. They would not create public confusion, but only public ridicule, soon followed by two new news directors.

On the other hand, if the NWS is scooped time after time, it gets nervous. It wants to eliminate competition. Then it will not get scooped. It will wait until the target touches the gun, then issue a warning, and presumably everybody will be satisfied. The NWS has been doing this for years.

Let's examine carefully that ambiguous word: "warning." If the observations of the NWS detect dangerous weather, of course it has a responsibility to issue a warning. Any independent forecaster will welcome that. But such a communication is a report, not a forecast.

Anytime the NWS labels something a warning it is a report of something already occurring in the vicinity. When the NWS says it is watching something closely for possible danger, remember they are doing what every observer is paid to do.

The watching the NWS does at WSFOs is of two kinds. One is to watch prognostic charts from the robot computer. The other is to watch hourly airway reports for existing weather conditions. Sometimes the "forecaster" looks out the window.

Round Table Discussion at Puget Sound Chapter of AMS

The Puget Sound Chapter of the AMS has been chartered since 1935. On November 6, 1985, it held a meeting on the subject of "Local Weather Forecasting — the Role of the Federal Government and the Private Sector."

The discussion was transcribed from a tape recorder. Its extemporaneous character reveals and preserves the attitudes and viewpoints of the speakers.

Harry Wappler from KIRO-TV was the moderator. Speakers appeared in alphabetical order. Transcriptions, excerpts, or summaries (possibly abbreviated) follow:

PHIL BREUSER, NORTHWEST WEATHERNET, P.O. BOX 855, ISSAQUAH, WA 98027-0855

(Phil presented his viewpoints of local forecasting in the private sector. When it comes to putting the right information into the right hands, at the right time, so people can make the right decisions at the right time, private forecasters are the best, and some of them are the worst. A local forecaster needs to consider (1) dollars and cents in his business, (2) communication with his client and also with other meteorologists, and (3) his display of confidence, without undue ego.)

JOEL CURTIS, FORECASTER, WSFO NWS, 7600 SAND POINT WAY NE, SEATTLE WA 98115

First, I'd like to say that I'm presenting the viewpoint of just the line person in the Weather Service, and some of these opinions are personal, and some of them are ones that I've gathered from working with my fellows. First, I'd like to talk about the role of the federal government, very briefly.

The federal government collects data, we run forecast models, we make general and a few specific forecasts, but the main thing that the federal government does, is **we protect lives and property** through our watches and warnings program. These are centralized functions that I think that best serve the public by a nationwide organization such as the NWS.

However, I want to say, that having talked to my peers at the Weather Service office in Seattle, that just about all of us agree, that the private sector has a vital role to play in this. And the way I see it, I think most of us agree, is that the private forecaster does provide those specific forecasts to the individual user that we either can't reach, or we are just not specific enough to provide for them.

And I really admire Phil's speech for bringing that out because that's really where we feel that we're at, with the private sector.

There's one source of contention between the private sector and just the people that work in the Weather Service, and that is, certain stategies of marketing the private local forecast.

One way to market this, I see, is that you can convince the user that you provide the essential specifics that he needs. And this we support wholeheartedly, because after all, this is the point that we believe that you're at.

But another way that you market sometimes, and it does really offer a lot of contention between us, and that is that your forecasts are so much better than the NWS forecasts, that the private forecast, because it's paid for, is superior to the free forecast. And I think this is misrepresenting things, and I think it's counter-productive, and **unethical**.

Now, my next point I'd like to do, is make a very brief comparison between the meteorological field and the medical profession. And in this case, what I'd like to do is use the patient as our forecast consumer. And the prognosis or diagnosis is really our local weather forecast.

First, some of the differences besides the obvious one, like it's a laboratory life science that the medical profession deals with. First of all, often the medical profession deals on an individual basis. This is true with the private sector.

In the public sector we're often reaching a group and it's rare that we get an opportunity to serve a private individual on a one-to-one basis.

The second difference is that in many cases the training is more intensive in the medical profession. We have all heard the moans and groans about our friends that are medical students that go to medical school.

Third, they have a very strong professional organization. I think we can all say it's much stronger than the American Meteorological Society. Money. Bucks. Power.

Also, in tune with this, they have very strict licensing. In order to practice medicine you must pass a board in each state. Also they have very little public contention within the profession. Rarely do you hear a doctor criticize publicly another doctor for a prognosis or a diagnosis of a patient.

And third, they make lots more money. What this really means is they are rewarded more by society for the services rendered. And some of the similarities that we've already talked about, first, private practice versus public practice, there's actually two sectors in the medical profession. Second of all, the responsibility is in varying degrees depending upon the situation.

In both fields we can have life or death situations, or protect tremendous amounts of money in the way of property. The other way, our field could also go down to somebody planning a garden party or something like that, as trivial as that.

Another similarity is culpability. Who has the responsibility for being right or wrong? Is it a legal responsibility, or is it just plain professional standards? And this is a question that we must ask ourselves as meteorologists all the time.

The last similarity is that there's a general assessment [?] versus specialties, such as the specialties private meteorologists provide. And then there's research and education.

Some of the questions that I feel can be raised by this comparison is: One, how does society view meteorologists versus physicians? In my own opinion, in our context of local forecasts, the public recognizes uncertainty in the weather, but will often not question doctor's orders.

So there's an inequity here between the medical profession, which is definitely an inexact science as well, and the meteorological profession, which we all know is an inexact science.

Second, can stricter licensing within our own profession increase public and private confidence in forecasting? A follow-up question would be, if a patient has several prognoses, i.e. forecasts, which treatment would he choose in the case of an emergency?

So this gets back to my contention that the weather service, in its fundamental mission **to protect lives and property**, provides a central focus in the emergency situation.

In closing, I'd like to say, and once again referring to my example, are government and private forecasters working with the same patient? I think generally, in focus of what Phil said, and what we believe in the Weather Service, they are not, often not.

And just as in medicine, there are rules for the specialist, which is analogous to the private forecaster, and certainly a role for the general practitioner, or the emergency room doctor, which is analogous to the Weather Service's general public **forecasts and watches and warnings**. Thank you.

RICHARD DOUGLAS, DEPUTY CHIEF, METEOROLOGICAL SERVICES DIVISION, WESTERN REGION HEADQUARTERS, NWS, P O BOX 11188, FEDERAL BUILDING, SALT LAKE CITY UT 84147

Thank you very much for your invitation to come speak tonight. Much of what I will touch on has already been touched on by Joel. In essence, we view the role of the National Weather Service, in regards to the role of the private sector, that there is a proper and necessary role for each. A strong public demand for accurate and timely weather information supports the need for both a strong public and a strong private sector for this type of information.

If one stops and considers this for a minute, if there was no need for any form of weather information, both sectors would be strongly in trouble, and in essence, down the tubes.

Thus a strong private sector supports a need, in my opinion, for the need of a strong public sector, and vice versa. In essence, we feel the private sector supplies that what the National Weather Service cannot provide. In reality, I do not see there is all that much competition between the two sectors, as might otherwise be expected.

I firmly believe that the primary National Weather Service

mission is its constitutionally mandated protection of life and property, and a provision for general forecasts and data acquisition to continue the primary responsibility of the National Weather Service. Thus I would like to give you a skeletal outline of the major program areas of the National Weather Service that we are currently engaged in.

In regards to data acquisition, I think everybody can agree, that before any forecast can be made, present weather must be determined. This is the primary goal of the National Weather Service data acquisition program. The major components of this data acquisition are surface and upper air observations, radar and satellite observations. The National Weather Service collects and distributes over 1000 land observations and 2000 ship reports each day.

In regard to radar, the National Weather Service operates 128 radars. Fifty-six of these radars are so-called network radars, while the others are local warning radars. The network radars are operated continuously, 24 hours a day, in good weather and bad weather, while local warning radars are operated as need be. Many of the radars are remoted to other locations, including locations used by the private sector.

The primary satellite system in operation is the geostationary orbiting environmental satellite, of which there are two primary satellites at this time, along with a number of polar orbiting satellites which collect numerous data.

We are also responsible, in my opinion, to provide numerical guidance, NMC models. We operate the National Meteorological Center In Washington, and I think, if we think of one point that maybe we can all agree on as to what the National Weather Service should do, it is that they should provide the observations and the numerical forecasts. I'm talking about hemispherical models operated and produced by the National Weather Service daily.

I cannot envision any private sector could hope to maintain a profit by maintaining, collecting, developing, and running on a large enough computer any numerical global product that would be of significant worth.

But our primary mission is that as a **protection of life and property**, and the provision for general forecasts. We issue watches and warnings to save life and property. This area receives our highest priority in terms of operation of the National Weather Service. Our warnings and severe weather watches are extremely carefully monitored.

The National Weather Service maintains a most comprehensive verification program. I would like to briefly show you how each and every National Weather Service watch or warning, along with every severe weather event, is verified.

Basically, we have a probability of detection that is the fraction of total events that were warned, compared to the total events that occur, for which a warning has been issued--a **false alarm ratio**. This measure, this false alarm ratio, is a measure of over-warning, it is the number of warnings unverified versus the number of warnings issued.

The critical Success Index, then, is the fraction of the time the severe weather was correctly forecast, when either the severe event occurred, and or, was forecast. The best score is 1.00, and the poorest score is zero. You see, if I attempt to measure [?] the skill of detecting severe weather, but we don't want to get into a business of over-warning. There is a necessary and proper balance between these two.

And finally, we measure the percent of warnings verified. This score can be written as the number of warnings verified divided by the total number of warnings issued. Our warning verification statistics are a matter of public record, as they should be.

To give you an example of what some of them can look like, this is the percentage and actual number of counties with and without an event while a flash flood warning was in effect during the year of 1984. As you can see in the Pacific Region, they were a perfect Region, because they only had two events, and fortunately captured both of them.

We do not pretend to be perfect. It is our goal, towards perfection. As you can see, in our Western Region, we verified about 75% of the warnings that we issued.

The National Weather Service also operates a public forecast system of generalized forecasts and extended outlooks. We believe these products provide the general public with current weather information and forecasts, especially through the news media.

These products oftentimes serve as a starting point for most interpretive and applied services provided by most private weather firms.

They are very broad and generalized in nature. For instance, in my home state of Utah, we subdivide the Beehive State into 11 broad climatic zones for which we issue a set of zone forecasts, which in essence are somewhat generalized.

For instance, for the Wasatch desert and the Wasatch Front of Northwest Utah, a typical forecast would read: For today, increasing clouds, local south winds 20-25 mph. For tonight and Tuesday, variable clouds with widely scattered showers and turning cooler.

We also, in addition, provide spot forecasts for a number of specific locations, but generally they are confined to the most important areas within a geographical area.

In addition, other major forecast programs of the NWS include aviation terminal and route forecasts. These forecasts provide for aviation weather information for safe and efficient flight operations through the United States and surrounding borders.

The National Weather Service issues 503 terminal forecasts and 327 route forecasts, which cover most of the important airways in the United States.

"In addition, the major National Weather Service office in Kansas City provides in-flight advisories to warn pilots of potentially hazardous weather.

We also have a comprehensive marine forecast program which increases the safety of life and property at sea. A typical marine forecast includes information for winds, wave, weather, storm surge, sea and swell forecasts. They are primarily divided into two sub-categories, one for the high seas, and marine forecasts for the coastal offshore waters.

We also maintain a fruit-frost and agricultural forecast to issue warnings of low temperatures injurious to sensitive crops, especially citrus and deciduous [?] fruits and tender vegetables

In addition, the National Weather Service is also engaged in fireweather forecasts. We provide warnings and applied forecasts and advisories to federal, state, and natural resource agencies in support of wildfire control and wildland management.

Most Importantly, however, we are engaged in disaster preparedness. We cooperate closely with federal, state, and local disaster agencies. Severe weather, in the form of hurricanes, tornadoes, floods, and other severe storms result in numerous fatalities each year, and in addition, cost 3.5 billion dollars. That is not a mistake, that is not a mispronunciation, it is 3.5 billion dollars.

The National Weather Service works closely with FEMA, the Federal Emergency Management Agency, and other state and local governments to develop a comprehensive disaster warning system.

Warning preparedness duties include recruiting severe storm spotters, working with the news media to insure rapid dissemination of warnings, develop back-up communication abilities, and provide technical assistance to local governments in developing emergency response and evacuation plans.

The dissemination of our products are mostly highly automated. We have the NOAA Weather Wire, Weather Radio, recorded telephone forecasts and observations. Most of the dissemination of these forecasts, is as I say, automated. That is to say, forecasters rarely give out forecasts, for the most part, to a specific individual on a one-to-one basis. We don't have the manpower.

The primary means of dissemination through the news media is the NOAA Weather Wire, and to the general public is NOAA Weather Radio. NOAA Weather Radio reaches a majority of users with 350 transmitters reaching 90% of the population. I reiterate, in general, our personal one-on-one contact is somewhat limited. Thank you very much, I appreciate your invitation for being here.

THOMAS J. LOCKHART, CERTIFIED CONSULTING METEOROLOGIST, CERTIFIED QUALITY ENGINEER (ASQC), METEOROLOGICAL STANDARDS INSTITUTE, P O BOX 26, FOX ISLAND WA 98333

(Excerpts) The private sector does not have a single voice. It is a whole bunch of different people. Some attempts have been made to get the private sector together, to have the benefit of a single voice. One such organization is the National Council of Industrial Meteorologists. [founded in 1968, address is 8801 Fox Drive, Suite 350, Thornton, Colorado 80221. Lockhart is a member and a past president]

You can't join them unless you work for a profit-making organization. The purpose is to develop positions in support of the private sector's interest in serving the field of meteorology. I have a draft of the NCIM position on specialized forecasting:

> Specialized forecasting was the first service of the private sector meteorologists. NCIM is confident when the NWS discontinues specialized forecasts such as fruit-frost service, the private sector will provide these needed services.
>
> NCIM feels that some weather-sensitive businesses such as those engaged in agriculture can maximize their benefits from a combination of information provided by an experienced forecaster, and from specialized measurements of local conditions.
>
> The private sector has the technical background to advise clients which measurements will be most important for their needs, and the knowledge of their business operation to recognize where weather information can provide greatest utility.

I am somewhat concerned with communication to the public. The public is not able to judge as we can the abilities of forecasters.

[Lockhart criticized claims of some forecasters who claimed insight to "achieve whatever 100% means."]

CLIFFORD F. MASS, ASSISTANT PROFESSOR, DEPT. OF ATMOSPHERIC SCIENCES, AK-40, UNIVERSITY OF WASHINGTON, SEATTLE, WA 98195

I'm neutral. (laughter) Private sector we will define as private forecasting firms, and also TV meteorologists. I will talk about what I consider to be the ideal situation, a few of the problems with the real world, and a little bit about the future. Obviously our goal is to give the best forecasts and nowcasts to the public.

I believe the National Weather Service has to serve as the backbone for this, because they have this tremendous investment in data collection and numerical weather prediction. The personnel they have is tremendous. It would be very difficult for a private organization to match these kinds of resources.

They have to serve, not only in data collection and numerical weather prediction areas, but also in the basic core of regional forecasting. It really has to be on the backs of the National Weather Service.

I think the National Weather Service should work also in trying to understand the local weather in an area, in other words a little bit in the research area too. That's something the Weather Service should be involved in.

Now the private sector, I think they are a crucial distributor of weather information, of forecasts, to the public. It's the media that really presents this information to the public. We're talking about TV, radio, and newspapers. In general, the newspapers are terrible, the radio is not much better, and the TV can be fairly good. It's that part of the private sector that has to do with communication.

The other thing the private sector should work on, is producing detailed forecasts, both temporally and spatially, for users. We're talking about bus companies, to transportation interests, to construction, this is something quite within the range of private companies.

Now the university role. We're neither federal nor private sector. We also have an important role: training people to go into the public and the private sectors, increasing the knowledge base, both on general forecasting and local forecasting, and also helping develop new technologies and techniques of forecasting.

So to do this, to bring the public the best possible forecasts, there has to be quite a close cooperative relationship between all these various sectors. And that would be truly an ideal world. And maybe the real world is not too bad. But I have noticed in the past, that there have been some tensions in the real world between these various sectors.

Specifically, it seems to me, more than once, there's almost a schizophrenic, or a love-hate relationship sometimes between the private sector and the National Weather Service.

Sometimes private meteorologists seem to feel that they are in competition with the National Weather Service, and with the need to prove to subscribers, or their customers, or their viewers, or whatever, they they can beat the Weather Service, in terms of forecasting.

I guess this is a very natural reaction that the private sector always has to fight against, because the Weather Service forecasts are free, and if the private sector is going to get paid to provide forecasts, they would have to do something better than the Weather Service to be worth their money. That's in the back of the minds of some people in the private sector.

These claims the private sector sometimes put out about superiority, and sometimes even direct criticisms, do prove unsettling to the Weather Service personnel. There are some very natural reactions on the part of the people in the NWS, that the private sector has certain advantages over them.

For instance, a forecaster goes on television. He might be 3 or 4 hours after the Weather Service had made their forecast, and so there is an implicit advantage there.

Or they have the advantage of having the National Weather Service's expertise and knowledge. They can read the discussion, and they're starting with that knowledge base when they make their own forecast.

So that's the hate side. I notice these tensions have been there. On the other hand, there's the love side of the relationship. Anyway you look at it, the National Weather Service and the private sector are tremendously dependent upon each other.

Even the largest private weather consulting companies depend upon the Weather Service for data, the basic forecast, the basic information. And the National Weather Service clearly depends on the private sector's distribution. There's no doubt about that either.

Another thing that would hopefully bind them together is their love of meteorology, which we all have, or we wouldn't be involved in this in the first place. These kinds of tensions we have to minimize as much as we possibly can.

Turning to the more local scene, I'm really hopeful that here at Seattle, and I think it's already starting to happen, that we can do something very very unique and something very very special in terms of bringing together the various sectors of the meteorological community, in terms of the local forecasting.

I think we can work together, first to try to understand the peculiarity of the local weather around here, and there certainly are a lot of strange local weather phenomena, I think together we can understand. I think together we can work on improving the local forecasting and nowcasting and distribution.

To give you one example, that we're working on at the U of W. We can try to create a local forecasting system, at the University of Washington, in which we bring in data into our computer, and then try to develop an expert system in which the forecasting knowledge that people in the private sector, and people in the Weather Service, have gained. That kind of knowledge can be put into software which helps guide a forecaster in making his local forecast decisions.

I can see the cooperation between various private, and the public, and the educational sectors has been increasing. I can only hope in the end we will be able to serve the public in a far better way because of this cooperation.

After the panel of speakers, discussion followed, condensed in part:

Phil Breuser

Private sector marketing — who is better? I do a lot of marketing in the private sector. People ask me, "Are you better than the National Weather Service?" I tell them, "How can you tell whether an orange or an apple is better?"

How can you tell whether 70% is better than 4 hours of rain in one day? Never is the marketing done on--"Who Is Better?" It's based on specific needs. It's apples and oranges as far as I can see.

Stanley W. Marczewski, Meteorologist, Member of National Weather Service Employees Organization, (Affiliated with AFL-CIO)

We're talking a great deal about privatization of the Weather Services under this administration. My question is directed to the private meteorologists. How will you feel when the administration starts charging you a user fee to get all of the data from the National Weather Service in the future? Will you pass this on to your customers?

Breuser

Gee, I don't know, I never thought about it.

Tom Lockhart

> As you probably know, in England, that's exactly what they
> do. The fundamental question is, whose data is it? Is it the
> National Weather Service's data, or does it belong to the
> people of the country?

Marczewski

> Right now we have a lot of cooperative weather observers
> across the nation supplying weather information. Down at
> the fruit-frost areas you have volunteer observers there.
> When you start charging, when you have a private
> meteorologist, they in turn won't provide that free service.
> They say, if you're charging for it (the forecast) you're gon-
> na have to pay for my observation.

Lockhart

> As far as the data itself is concerned, its difficult for me to
> understand how you could put a user fee on that.

Marczewski

> You say that the taxpayers own the observations. Well,
> we're paying for them right now. Then why should we go out
> and pay double? We already paid for the observations. Why
> should we pay you a profit to supply us with a forecast?

Lockhart

> A forecast is not an observation. A forecast has to be based
> on some observations. Are you suggesting that the private
> sector, in order to provide forecasts, would have to have
> their own data network around the United States?

Marczewski

> Right! [derision from audience, then a question, voice too
> faint]

Richard Douglas

> We feel there are some things inherently governmental.
> One of them is the primary mission of the National Weather
> Service is to issue warnings.
>
> To my knowledge, a private meteorologist can issue a
> warning if he so chooses. I'm pretty sure that's true. He can
> do it legally. We are strongly against that. Let me give you a
> specific example.
>
> In my home town of Salt Lake City we have 3 TV sta-
> tions. How confusing would it be, when you include the Na-
> tional Weather Service being a 4th source of information?

Some people issue warnings. Some people don't. Who do you believe? Is there a warning issued, or is there not? We believe the warning belongs on this side of the fence, which is the National Weather Service.

What we've seen here today, are some things that are clearly inherently governmental, and then there are some things that are clearly what the private sector should be writing [?] and there is a gray area in between.

Lockhart

I agree with you completely. At the annual meeting [of the AMS, January 1984] the subject of warnings by the private sector versus the federal government was the subject of a fairly prestigious group.

The position was that the federal government had the responsibility to make these warnings, and the private sector was fully in accord with that. The confusion element of having everybody putting their two-cents-worth in, is so horrible, I don't think there's any problem on a consensus on that subject.

Joel Curtis

One quick thing I would like to learn from the private sector. You're out there getting all that experience with these individual users, and things like this. Because of that, you're tailoring your forecasts so specifically, I think we in the Weather Service could learn a lot from you, if you could tell us the **secrets** of being able-- (laughter). If you're making money, you must!

Breuser

Come on, Joel, do you think I would do that? More communication, that sounds good. On the other hand, if the National Weather Service started defining their forecasts more, I would be out of business! Because the ability is there at the National Weather Service. You see, your mandate is to serve the general public, and I'm not gonna even touch that, because there is no way I would even come close to that. So maybe I might have a few tricks to the trade, I really need my living too.

Lockhart

I think it's an apples and oranges thing you're talking about. It's not a matter of capability in forecasting. I think you could both make the same kind of forecasts. It's a focussed forecast, for a particular purpose, for a particular place, at a particular time.

And the National Weather Service can't put time into doing that kind of thing, shouldn't put time in to do that kind of thing. And there is a need by the user, by the construction company, or the movie studio, or whoever it is, for that specific information. It's the same forecast, but it's just a different focus.

The preceding pages are an exhibit of the members of the meteorological profession talking to themselves, which is one of their handicaps.

The reader should know by now there are two separate private sectors — the older familiar non-federal forecasters, few in number, serving a small group of clients with "tailored" products — and the newly designated private sector of media "disseminators."

The federal bureaucracy tolerates, and pretends to warmly support, the older group (such as Lockhart and Breuser), as long as it does not intrude into the turf of public forecasting. This older group either accepts its share of the pie without question, or prudently pretends to do so.

The new group of media disseminators, who either are not meteorologists by any fair definition, or have limited qualifications (such as most of those from short service in the military, but there are a few exceptions) are grateful for the artificial prestige bestowed by the federal bureaucracy, and the captive American Meteorological Society. They are partners, and enjoy benefits.

I no longer strive to persuade these groups. This book is for serious forecast users, and it strives to explain **Why Can't They Get It Right?**

Serious forecasts users need not stand in awe of these meteorologists of assorted qualifications and motivations. They need to adopt an attitude of evaluation, to kick the tires, and look under the hood.

The tools and techniques are available. SFUs must offer a fair reward to real forecasters who also use intellectual subjective decision-making skills.

Golden Phrases to Remember

Agenda for Action, American Mcteorological Society:
QUICK RESPONSE FORECASTS AND WARN-
INGS   WARNINGS OF SEVERE WEATHER PREVENT
LOSS OF LIFE AND PROPERTY   WE NOW KNOW HOW
TO DESIGN EXPERIMENTS TO FURTHER IMPROVE WAR-
NINGS   WE CAN NOW **PINPOINT** DESTRUCTIVE
PHENOMENA MORE ACCURATELY   TECHNOLOGY IS
AVAILABLE TO COMMENCE **PROCUREMENT** OF
SYSTEMS   GREAT COMPUTING POWER WILL
DEVELOP FORECAST PROCEDURES   THE PRIVATE
SECTOR SERVES **SPECIFIC USERS**

Joel Curtis:
WE **PROTECT LIVES AND PROPERTY** THROUGH OUR
WATCHES AND WARNINGS PROGRAM   CONTEN-
TION THAT THE PRIVATE FORECAST IS SUPERIOR IS
UNETHICAL

Richard Douglas:
PUBLIC DEMAND FOR ACCURATE AND TIMELY
WEATHER **INFORMATION**   THE PRIVATE SECTOR
SUPPLIES THAT WHAT THE NWS CANNOT PROVIDE
THE PRIMARY NWS MISSION IS ITS **CONSTITUTIONAL-
LY MANDATED** PROTECTION OF LIFE AND PROPER-
TY   WE ISSUE **WATCHES AND WARNINGS** TO
SAVE LIFE AND PROPERTY. OUR WARNINGS AND
WATCHES ARE EXTREMELY CAREFULLY
MONITORED   THERE ARE SOME THINGS IN-
HERENTLY GOVERNMENTAL. ONE IS TO ISSUE
WARNINGS.

These golden phrases reveal a unity of propaganda theme, from a tarnished scientific society, a GM-15 grade manager from the Western Region of NWS, and a local NWS forecaster in Seattle.

Occasionally the important word "forecasting" appears, but the word "warning" is preferred, and often linked with "timely." The public demands accurate "information" rather than accurate forecasts. Destructive weather is pinpointed rather than predicted. The "watches" are carefully monitored. Forecasting is a form of electronic technology, dependent on the purchase of computers.

These are the viewpoints of a weather observer, not a weather forecaster.

When the Weather Bureau was created in 1890 one of the noble objectives was to save life and property, which implies prediction, so preparations for severe storms can be made in advance. But such legislation does not appear in the Constitution.

And why is the issuance of forecasts, which clearly requires subjective intellectual ability, inherently a governmental function?

If, as Douglas claims, the private sector can provide what the government cannot provide (presumably "specific" forecasts), why should not the public enjoy such benefits, especially after the public pays the bill for all that **data acquisition** and computerized guidance?

Why do these peculiar words "specific" and "specialized" keep cropping up? Maybe the November 13, 1981, windstorm provides the clues. The Mitchell Award forecasters made a "general" forecast for winds 35 to 45 with gusts to 55· mph for Portland, Oregon, and William Wantz and Charlie Feris made a "specific" forecast for gusts to 70 mph (4 1/2 hours earlier, after holding off for about 2 hours).

In Minneapolis, but only about 2 hours before the general forecast, Chuck Wiese made a "specific" forecast of gusts to 90 mph (a bull's eye).

Wantz, Feris, and Wiese understood the mechanics of a Columbus Day Storm. Of course, no one ever explained that to the computer in Washington, D. C.

When Chuck Wiese nominated Wantz and Feris for a Mitchell Award, someone threatened him over the telephone.

I can almost hear a national chant, a chorus from 50,000,000 serious-forecast-users, **Why CAN'T? They GET? It RIGHT?**

The Ideal World of Professor Clifford Mass

Professor Clifford Mass says the National Weather Service has to serve as the backbone for the best forecasts and nowcasts because it has this tremendous investment in personnel and data collection and numerical weather prediction. Whose investment?

The vested interest of the National Weather Service seems primarily to be its self interest, closely related to a cylindrical administrative structure.

He says the private sector (the older, forecasting sector, not the newer dissemination sector) should work on producing detailed forecasts for users such as bus companies, transportation interests, construction, etc. Such detailed forecasts are within the range (of expertise) of private (independent) forecasters.

Why not expand the list to include farmers, office and factory workers, school children, sports activists, truckers, aviators, vacationists, and others in our busy nation? Well, the NWS needs to find something glamorous for all those WSFO forecasters (?) to do, like save lives and property. Also there is need to maintain the ideal world of academia, which means to keep the grant money flowing.

The following counter-proposal may surprise Professor Mass. Chuck Wiese and I also would like to keep grant money flowing, providing it is used to improve weather forecasting. And even more grant money will be available if we get rid of the watchers and warners at the 50 or so WSFOs.

If academia will help the best interests of serious forecast users, the SFUs will persuade Congress to help academia. The value of scientific research is well proven. Academia ought not need to kowtow to the self-interest of bureaucrats in the National Weather Service.

What has mutual love of meteorology got to do with it? Maybe Professor Mass means satisfaction from scientific achievement. That is in short supply today in weather forecasting. Even the American Meteorological Society admits the atmospheric educational enterprise is encountering significant problems.

But the problems don't stem from the technological revolution, as Dr. Weinstein et al claimed. They stem from subsitution of greed for scientific achievement, and from the surplus of third-rate minds who have seized positions of leadership.

Robin Cody's article appeared in The Oregonian in late July 1985. This discussion by the Puget Sound Chapter was in early November. Professor Mass noted that claims about superiority, and sometimes direct criticism, proved unsettling to National Weather Service personnel. The harassment of

Chuck Wiese, independent forecaster, is also unsettling.

Mass describes the "implicit advantage" of a TV forecaster (who else is there in Seattle except those at KING-TV?) in making a forecast after the NWS made its forecast, thereby having the advantage of NWS expertise and knowledge.

Well now, if an independent forecaster is competent enough to recognize an inaccurate forecast from the NWS, more power to him.

And then the NWS can watch its TV screen, and take advantage of KING's expertise and knowledge, and revise its forecast (which it has done several times). The public ought to benefit by this healthy intellectual competition.

Professor Mass seemed to overlook that he was speaking into my tape recorder, in addition to addressing an audience sympathetic to his views. He was the president of the Puget Sound Chapter. In May 1986, Joel Curtis was elected president.

And this is not all. Like the spieler in the television advertisement for the vegetable slicer, there is **more**, about the public-private partnership in the new group, the media disseminators.

Is the Government in Collusion with the Media Disseminators?

Robin Cody quoted me, "Government, in collusion with the media, has lowered expectations . . . people don't realize what they are missing."

I have since reviewed the definition of collusion. That means (American Heritage dictionary) a secret agreement for a deceitful or fraudulent purpose. That is not far off. Maybe intrigue would have been a better word, to imply selfish, petty actions rather than criminal ends.

Portland has a television station KATU-TV (ABC) which has a durable weather reporter named Jim Bosley. He gained an AMS Seal of Approval some years ago although as he explained to the Oregon Chapter at a meeting, he was really a stand-up comedian.

Also, he is a skilled master of ceremonies, much more adept with language than the put-on verbal clumsiness he uses on his weather show to attract his targeted audience.

Everybody likes Jim Bosley, including me. Also, Jim is more sensitive to the needs of the forecast users than some other weather-reporters.

Bosley also is a co-host of a morning talk show for housewives, AM Northwest. Three times in a period of 13 months, Bosley had George Miller and Phil Peck from the NWS as guests on his AM Northwest show. By unplanned coincidence, the second one came two days after Robin Cody's article, Why Can't They Get It Right?

Bosley, Miller, and Peck were disturbed. Margie Boulet was co-host.

Bosley

A little rain this morning. Boy! There were some guys on television that didn't say it was gonna rain a little this morning. Ha ha! There's a little bit of cloudiness causing some dribble-- dribble, drizzle. I love that term. Whether we will remain a 53rd dry day in a row, or not, is debatable. I suspect, in talking with the gentlemen with the NWS that are here--they think maybe we may get a hundredth of an inch, which would break this streak of dry days we've had.

Boulet

What do you think? Because I know you do your own forecasting.

Bosley

Right, right. I think it might. We are in a drought here. We have a weather test coming up. We're gonna talk about some interesting things today, and show you some clouds. Is that partly cloudy or mostly cloudy?

[Bosley holds up Cody's article] This thing was written by Robin Cody. Remember Bob Lynott, he used to be on television years ago? You're too young. Bob's upset now. That the National Weather Service is not doing a good job in forecasting. He says that the government is in collusion with the media. That television forecasters are simply repeating what the Weather Service says.

In the case of John Walls at Channel 6 that's true. Uh, Jim Little, I'm not sure. Dave Apple [also on KATU-TV]? No, he makes up his own forecasts. Here at Channel 2, I do my own, right or wrong.

[Bosley introduced Miller and Peck] This article really beat up on you guys, I mean it **really** beat up on you guys. Uh, was it a fair article?

Miller

No. The article was slanted towards those particular individuals that you had mentioned, and the fact that some of them have worked for the Weather Service in the past — When I talked with Mr. Cody, some of the things that I thought would be out in the article were not brought out.

Bosley

Phil, did he interview you? Was it all accurate?

Peck

Yes, I spent about an hour. It was no pretense at balanced reporting.

Bosley

Lynott is taking you to task. Why is he doing that?

Miller

He believes our forecasts are too conservative, that we do not forecast the change. I'd like to take today's forecast, for instance. Two days ago [Sunday] we were forecasting temperatures that were in the 90s at that particular time, and dropping them down into the 70s, and I don't call that conservative.

This was 9 a.m. Tuesday, July 30, 1985. Portland's high was 85 on Monday, with no rain for 52 consecutive days. Fire danger was very high. The extended outlook issued Sunday evening, published in *The Oregonian* Monday morning, called for a few showers or possible thundershowers Wednesday in Western Oregon with highs 75 to 85.

The forecast issued Monday evening, published Tuesday morning, predicted Portland's high near 75 (it was 66), with partly sunny skies, chance of showers 40% for Wednesday. As the three forecasters (?) were talking, a pronounced change was already under way. But they seemed unaware of its magnitude.

Heavy rain had already swept across parts of southeastern Oregon late Monday helping firefighters to combat dozens of new lightning blazes on forest and range land. The Forest Service reported 1,300 lightning strikes in a 2-hour period Monday afternoon. This discussion was the morning afterward.

Bosley

Did you, did you say it was gonna rain this morning?

Miller

Wednesday was the first time, I believe, that we had rain in the forecast. That was in the extended forecast.

Bosley

I didn't say it was gonna rain this morning. So, would, that make us wrong? I mean, was the forecast I issued at 5 p.m. [yesterday] and the one you issued at 3 p.m. yesterday, were they wrong?

Miller

I think there was a 30% chance of showers.

Peck

Yah, chance, small, slight chance. [published forecast, made about 8 p.m. predicted chance of showers, no percentage given, for Tuesday night.]

By 5 p.m. that day Portland measured .08 of rain; Salem had .25 with a high of only 62 degrees. Burns in eastern Oregon measured .88 which flooded drains, but that actually fell late Monday, prior to the TV discussion. Eugene, in the center of western Oregon, measured .55 with a high of 62.

Boulet

[Is it] more difficult to do weather forecasting in the Northwest?

Peck

It is, it is difficult in that, uh, uh, we're like any other business, and, uh, we're dependent upon information, to make decisions based on incomplete information. The Pacific Ocean is a sparse data area. We don't know a great deal about, uh, what's happening out there sometimes.

Bosley

Okay, let me ask you a question. Are we in collusion? Is Channel 2 in collusion with the National Weather — Do we have an agreement that we will only use your forecasts?

Peck

Not that I know of. No, indeed.

Miller

No sir.

Bosley

Okay. John Walls mentioned in here that we meet every three months for dinner.

Peck

I don't think there's any sinister imputation —

Bosley

I don't either, but the way it was written is that we're all in bed together, see, and whatever you say, I'm gonna say. Well, there's nothing like that, is there?

Peck

Of course not.

Miller

None whatsoever. Our dinner meetings that we've had, have only been to exchange information as far as what the Channel is doing, and what we're doing, as far as the advances —

Boulet

And have fun! I've heard about those dinners.

Bosley

We just sit and complain about the way things go on, that's all.

Miller

We compare notes.

When I naively recorded that, I thought the reference was to the 3-times-a-year meetings of the Oregon Chapter of AMS, which are mainly social meetings including spouses and guests. To my amusement, in February 1986, George Miller distributed a message to the media disseminators:

> It's been almost a year since we got together as meteorologists and/or TV weather personalities. I have reserved the small room at the [Italian restaurant] from 6 to 8 p.m. Wednesday, February 19th. Please attend and perhaps say "Good Bye" to Dr. John. Call me if you can't possibly make it.

I was once a TV "personality." (?) If I had been invited, I would have offered to buy a round of drinks.

On September 2, 1986 I called George Miller. The grapevine said Miller had called another meeting with the media disseminators. I would like to attend. I would not disrupt the meeting. I would be quiet and listen. Where and when would it be? Miller said he would "prefer not to." Was it a closed meeting?

No. But Miller wouldn't tell me the place or time. But if I could find out, I was welcome. Miller was curious. Why did I want to attend? Because Miller's influence on the media was one of my topics. Miller offered a compromise. He would query the others. He would explore their feelings.

A week later I called again. Miller said those he did contact declared if I showed up, they would not come. Miller would not reveal the place.

The bureaucratic public-private partnership met September 10. In the small back room of the Italian restaurant. I arrived after they had ordered pizza and beer.

They saw I intended to stay John Walls, Jim Bosley, and Phil Peck departed. Dave Apple arrived after 7 p.m. His show is 6:30 to 7:00 p.m.

The agenda was disrupted. A reporter was present. Some things are not for you to know. Matters in the public-private partnership.

George Miller, Jim Little, Miles Muzio, and Dave Apple hung around. For about an hour. Nothing much was discussed. The beer and pizza tasted good.

In the next chapter we will explore what is going on in television weather. Or rather what should be, but is not.

The public-private partnership is a myth, asserted and enforced by an arrogant bureaucracy.

The private sector now has two distinct segments — those forecasters serving private clients for a fee, and the newly designated media weathercasters.

The wise heads in the AMS need to shake the bureaucratic monkeys off the back of that scientific organization, and steer it back into an intellectual course. Every scientific-minded member of the AMS should be ashamed.

Unstated is the autocratic determination by the government that independent "forecasting" services will not have access to the mass market, the exclusive sales territory of the bureaucracy, the gold mine which provides justification for its cylindrical administrative structure.

Forecasting requires only a few hundred special-type minds to serve the entire nation.

Serious forecast users must restore "man" to the "man-machine mix."

Hardly anyone listens to those pitiful figures who proclaim the world is coming to an end tomorrow.

"The primary NWS mission is its constitutionally mandated protection of life and property." (Richard Douglas, Deputy Chief, Meteorological Services Division, Western Region Headquarters, NWS, Salt Lake City)

Academia ought not to have to kowtow to the self-interest of bureaucrats in the NWS.

Problems don't stem from the technological revolution. They stem from the substitution of greed for scientific achievement, and from the surplus of third-rate minds who have seized positions of leadership.

10

Television —
Gridlock or Gold Mine?

Television is the tabernacle of the 20th century, or perhaps the Mt. Everest, or the Communication Breakthrough surpassing the invention of the printing press. If it has not yet "found itself," that is our collective fault. We are only concerned here with weather forecasting.

I gained some attention when the local news program was 15 minutes long, 4 minutes each for Weather, Sports, and News, in that order, with a 1-minute commercial between each. It was assumed that Weather would encourage the tune-in.

Now is seems TV management is worried about Weather tune-out, and with good reason. The NWS doesn't know how to make genuine forecasts. The public and TV management often wonders about their inaccuracy. Public service for the **Weather Tomorrow** is stuck in a traffic jam.

Like a curmudgeonly cranky cop in a grade B movie, I am trying to get traffic moving again. I am appealing to those who really have to get to work, the serious forecast users (not the ones who have engineered weather out of their lives) to organize their political clout. We need Walt Disney to make another movie, this time a comedy about the National Weather Service.

TV management ought to realize they don't need a Willard Scott or a Jim Bosley for comic relief on grim news programs. They can just point to the forecasts of the NWS.

The NWS is maintaining a monopoly. The handful of independent forecasters ought to get together and sue the federal government for anti-trust violation.

I was a failure from the standpoint of show-biz on TV. I was too square, socially awkward, orally hesitant, lacking in wit, homely, and with a voice that caused a good instructor to throw up her hands. And I knew nothing about coping with TV management.

I had no delusions of grandeur. My television title was "Mr. Weatherman." I never owned a car until age 34. I rode a single-speed bicycle for a total of 26,000 miles. As I told the Oregon Chapter, I have no position, no authority.

But I do know how to read weather maps. They are easier than everyone thinks. Some people in the Pacific Northwest found that out in the 1950s by watching me for 20 minutes each week.

The self-depreciation above doesn't mean a lack of self-confidence. I am only pointing to characteristics which are unimportant, in order to distance myself from show-biz types.

I am self-confident about predicting the weather for tomorrow. Isn't that the only thing you need from a forecaster?

The reason so many TV weather reporters try to be funny is because they have little else to offer. Aren't you tired of all those endless temperatures?

Weather Maps, the Road Maps to Tomorrow, the Undiscovered Gold Mine

The subject of weather maps and their use is known as synoptic meteorology. The patterns are a synopsis of weather displayed geographically. The patterns move, and the maps are like snapshots.

The technique flourished during World War II. It inspired Jack Capell and me (and later Chuck Wiese). For some dark reason (maybe because it requires subjective intellectual ability) it has been forgotten in the dusty storage room. Perhaps we should say, relegated to the electronic robot.

Those gaudy charts and satellite pictures you see on TV are not weather maps. Those are "visual aids" dreamed up by news directors and show biz personalities. They don't understand weather maps, and hence can't explain them to the public. Meanwhile, the public "wants what it is getting."

Yes, there are some minor details such as weather codes

and map symbols on real weather maps that seem complicated (but are not). The essential features are easy to learn. I plan to write a slim book on the subject, understandable to anyone who can read at the 9th grade level.

Then serious forecast users should demand a few minutes each day on at least one TV station in each community, to look at these maps. If the TV industry can only spare 2 minutes, we can record it on our VCRs, and study it carefully.

With a modest amount of instruction and a little experience, the average television viewer can then make a homemade forecast with more accuracy than the NWS. Then after the NWS completes its **watching** for a significant event, and comes out with a **warning**, the stay-at-home forecaster can cackle. Did you know that cackling is one of the big pleasures in life?

I am not only a kibitzer. I am a cackling curmudgeonly crank.

There is nothing remarkable that Chuck Wiese came to work in Minneapolis on Friday afternoon November 13, 1981, and soon said "My God, I gotta call Portland." Or that Bill Wantz and Charlie Feris couldn't wait beyond 1 p.m. to notify the electric utility crews (4 1/2 hours before the NWS notified the public with its Mitchell Award forecast of 35-45-G60, and 5 hours before the revised forecast of G55). All we needed to know was G90. The weather map was staring them in the face, and scaring them.

People love to gamble. Weather forecasting is more fun than lotteries or slot machines. The stakes are high and the odds are more favorable. Almost anybody can win, often. That is, if you use a little reasoning. I am not interested in lotteries or slot machines. I was a professional gambler for 25 years.

What did I win? Well, not much money. But here is a story to illustrate. I don't even remember the specific instance. This is Howard Graham's story.

Howard was out on a forest fire for the U. S. Forest Service. I was in the fire weather office in Portland. I was predicting substantial rain for the next day. The Weather Bureau was predicting continued hot and dry weather.

Howard called me on short wave radio. Was I really sure? Very sure. Okay, then we will take it easy tonight. The next day they broke fire camp, but they didn't move out, because they were stuck in the mud.

Someday soon the public will discover that the guessing game about the weather tomorrow is just as interesting as baseball and football scores. And quite a few weather junkies will collect their spending money from their less attentive buddies.

Then the television program managers will wake up to realize they have a new advertising gold mine. (Please send me a small commission). Of course, they will have to pay some independent forecaster for the forecasts that now come free over the NOAA Weather Wire (at exhorbitant expense to us taxpayers). But that will be easy. Just cut the salaries of their Chief Meteorologists.

What if the Chiefs can't explain the weather maps? That's also easy. Turn it over to someone on the news staff. They soon can learn. Most of them are versatile. Give them a copy of my next book on weather patterns.

What about those who have engineered weather out of their lives? Well, the program managers can keep the personality figures like Jim Little, who invented that expression, and who claims he has "a constituency of maybe 100,000 people every night." Serious forecast users will only ask for a small piece of the time-schedule pie.

The Heavy Hitters of Portland, in Weather

When Robin Cody used the phrase, "the heavy hitters of Portland" (in weather) he meant the weathercasters on the network stations during prime time on weekdays. That meant John Walls, Jim Bosley, his colleague Dave Apple, and Jim Little. Walls has left TV, but is heard on KXL radio.

Cody was not talking about home runs. He meant where the TV program directors have placed their bets, on the weathercasters who can make the NWS products shine the most. I doubt if anyone ever told them there are better products, certainly not the NWS.

The reader should understand I am not trying to banish radio and TV weathercasters. I am urging them to disseminate genuine forecasts, and to learn how to explain the basis for those forecasts. The NWS is selling them short.

It is beyond my scope to describe TV weathercasting across the nation, but comments from travelers are consistent in rating nearly all TV weather programs even below those in Portland, Oregon. I am now convinced that Jack Capell and I set standards in the 1950s, with vestiges that remain today. It was not that we were brilliant. We were weather-map forecasters, not "graphic-aid reporters."

Today, with two 3 1/2-minute shows per week, and almost totally paralyzed, Capell offers the best weather program in the Portland area.

Dave Apple (KATU-TV) at one time made radiosonde observations for the Weather Bureau. He ignored my inquiry about his résumé. He features an occasional partner, Bob the weather cat.

Jim Bosley has been weathercasting on KATU-TV since 1975. He told Bob Beck, of *The Columbian*, Vancouver, Washington, "We all look at the same satellite photos and the same weather maps. The only thing any of us really knows for sure is what the weather is doing right at the moment."

Jann Mitchell, writing in *Northwest Magazine* of *The Oregonian*, for Sunday, October 10, 1982, quoted Jim Bosley, "People say they watch me because the rest of the news is so depressing. If you smile and your bald head shines and you act like you're up, then they feel better."

She also wrote, "He claimed he once made a couple feel so much better that they conceived a son during his forecast and named the boy after him."

As this book says, forecasting is lots of fun.

It is too early to judge Miles Muzio, the successor to John Walls. A new man has to be prudent, under existing circumstances at least.

Jim Little is a Portland boy, a graduate in mathematics from Linfield College in McMinnville, and was a Weather Officer in the U.S. Air Force for 5 years.

He has been Chief Meteorologist at KGW-TV since late 1979. Within a year he was on the Executive Council of the Oregon Chapter of the AMS, and served as president in 1984.

Jim Little concentrates on the entertainment side of weather, his personality, the picture show of satellite scenes, color graphics, and endless recitations of temperature data.

I used to live in central Oregon, in a windswept area. One evening during a high-wind episode, I tuned in for Little's weather show. It started off with first things first, this time the "toilet paper caper."

He had encouraged young students to set out a piece of toilet paper for a short period, maybe it was 5 minutes, to count how many raindrops fell on it. By that time I had lost interest in the rest of the show.

I have appreciated one thing. Jim usually shows the pattern of upper winds at 18,000 feet. However, he often displays the chart for about 2 seconds, and rarely explains significant features of the pattern which relate to the forecast problem.

This ought to a major part of the presentation, at least when important changes are impending. Apparently, Little uses this chart for window dressing, a sort of technical decoration. If I can't grasp it, how can the average viewer?

At the special meeting on Professionalism, Jim offered this view of science:

> I'd like to say something that I consider kind of an essential element of science. We did a thing on Channel 8 several years ago, called Nature's Weather Forecaster. We asked people to send in things that they observed about nature that helped them forecast the weather, wooly caterpillars and so forth.
>
> The big common postcard was people look at their cats. And when their cat got frisky and laid on its back, and so forth, it rained invariably within 24 hours. This is a good thing. Very accurate and so forth.
>
> But I always said "What happens when that cat dies? Can you be sure the next cat you get does the same thing?" And you can't. What you do is get a barometer, and when your barometer breaks you get a new barometer. Well, isn't it the same thing when you die?
>
> That's why we have computer models. I think that's the essence of science, an essential element of science is to **quantify** what's going on, report it.
>
> That's part of the scientific method, making that report, and going on to the next thing.

An article in The Oregonian August 27, 1986, by William R. Greer of the New York Times News Service, said:

"Sociologists say that *lite*, which started as a marketing term used to denote dietetic products, has become a metaphor for what Americans are seeking in their lives."

That is lite, as in less filling, fewer calories. For the most part, in the Pacific Northwest and elsewhere, weather forecasting on television is pushed aside in favor of lite entertainment.

Jack and Sylvia Capell, Weather Forecasters

There are two others in Portland who deserve honor beyond my limited ability to bestow. They labor on weekends at KGW-TV.

Jack and Sylvia Capell are weather forecasters. They also are saints. Sylvia doesn't have a degree in meteorology, but her hands and feet are extensions of Jack's, and her mind is in synch with his. In marriage ceremonies, the minister (or priest or rabbi) often refers to two individuals blending as one. This is a heart-warming example (also tearful).

Peter Farrell has granted permission to reprint his tribute as follows:

(From "Behind the Mike" March 3, 1983, *The Oregonian*)

K G W SHOW PROFILES COURAGE OF CAPELL

Legs and arms paralyzed, he arrives at the KGW-TV (8) news set in a wheelchair. He studies the weather maps and data that his wife has collected from the wire service transmitters, as she does every day. [In 1983 Jack and Sylvia did the noon weather show each weekday, but not on weekends.] And when the newscast is under way, he gives Portland the latest weather information, as he has done since KGW's first news broadcast in 1956.

He is Jack Capell, KGW's noon news meteorologist, a determined man who for 11 years [In 1983] has been battling amyotrophic lateral sclerosis, a degenerative illness often referred to as Lou Gehrig's disease. [Apparently Jack is a rare, non-classic case.]

Capell, a man with a fierce will to live, is the subject of a "P. M. Magazine" profile at 7:30 p.m. Thursday on KGW. In this provocative — if woefully short — six minute piece, "P. M." host Paul Linnman reports on the 59-year-old weatherman and "the courage it takes for him to live his life."

A former hockey player [who helped found Portland's Amateur Hockey Association], Capell swims nearly every day [not any more] to stay as fit as possible. It is not easy for Capell; his labored strokes barely keep him afloat.

On the job, his wife of 31 years, Sylvia, does the physical work that he cannot. A meteorologist from way back, Jack Capell analyzes the satellite photos, and he tells noon news viewers of the trends and temperatures.

The "P. M." feature on Capell isn't designed for sympathy, though it's sure to inspire. Linnman says his purpose was "to tell a story that's been sitting under our noses at the station for a long time . . . Here's a guy viewers have been watching for years, but how many people know what he goes through?"

Those who watch Linnman's feature will come to know Capell as a man who doesn't take much for granted. "How can a guy ever be unhappy when he can walk?" Capell comments in a reflective moment. "I didn't realize how much it meant until I started to lose (the ability to walk)."

Linnman, a modest, affable fellow who doesn't often brag about his work, and who refers to his formula fluff program as "disco television," is especially proud of his Capell profile. "It seems to me if 'P.M.' can do anything, it's to inspire," says Linnman. "I defy anybody to watch Jack Capell on the air after they see this and not think about his life, about what it takes for him to do his job."

This author has done much thinking about the life of Jack Capell, and his wife Sylvia. It is now November 1986, and in December Capell will complete 30 years as a television weather **forecaster**, and not merely a media disseminator.

In late 1971 when Capell was stricken, KGW-TV and Jack went through a short period of uncertainty about his future plans. The station has been extremely cooperative in assisting Jack and Sylvia.

The weekend shows at 5 p.m. are only for 30 minutes, and Jack's time is short. But he displays both surface and upper air maps, as a real forecaster should, and as every serious forecast user wants to see. They understand why he and Sylvia usually Get It Right. Jack and Chuck Wiese sometimes confer on the sister-station phone line, because both are on duty on weekends.

Chuck was only 12 when this author left television. Chuck knew he wanted to be a real weather forecaster after watching Jack for years (and pestering him). Jack is a patient soul.

(Note: This topic was written in November 1986. It is now early March 1987. Sylvia suffered a moderate stroke, requiring hospitalization. She is home now, slowly recovering, but unable to care for Jack. Their oldest son, John, has taken over his mother's strenuous duties. Young Tom is away at

*college, but assists during visits. Earlier, he was Sylvia's
assistant.)*

LIFE IS SO DAMN UNFAIR.

Seal of Approval of the American Meteorological Society

About 1960, because many weather programs on TV
were degrading the budding technology of meteorology, the
AMS began an effort to upgrade such presentations. The purpose was to recognize and support creditable performers, and
incidentally, promote the AMS and the "advancement of its
professional ideals."

It is unfortunate that those ideals have been cheapened
by the shoddy self-interest of the National Weather Service.
The Seal of Approval Program is in shambles.

Jack Capell was granted Seal #19 in April 1960, and I got
#39 in May 1962. In Portland most of the weathercasters have
the Seal, but rarely if ever does anyone display it. What competitive advantage is there?

The history of the Seal program is hazy, but in my opinion it goes something like this. At first the NWS was slow to
recognize the enormous public relation opportunity of television. Belatedly it saw that mavericks like Capell and Lynott
(who disagreed with the official forecast when necessary)
would hurt the government monopoly on public weather
forecasting.

The NWS encouraged the AMS to relax its standards for
approval, during which many non-meteorologists, like Jim
Bosley (a stand-up comedian), were granted Seals of Approval. These media disseminators were artificially promoted
as professional cooperators of the NWS, the touted public-
private partnership, as already described.

Recently (apparently to protect the turf of the established
Seal holders), the AMS has revised upward the requirements
for a Seal. The applicant must be a full member (not merely
an associate or student member) of the AMS, which generally
means a degree in meteorology.

If I were still doing weather on TV, I wouldn't bother to
display the Seal either. In practice, because of the variable
rules, and mainly because of the domination of the NWS over
anyone in television, the public doesn't understand the mean-

ing, or the lack of meaning, of a Seal of Approval.

This takes us back to an elementary rule of life. Quality can best be determined by the consumer, by looking under the hood and kicking the tires. Such will be one of the incessant activities of the Weather Tomorrow Society.

Soon the quality of public weather forecasts will rise significantly, and taxpayers will save money instead of spending more money for more computers and more additions to the administrative cylinder.

The Research Project of Professor Driscoll, Funded by NASA, an Example.

As a life-long copycat, I have learned how to write titles from Dr. Allan Murphy at Oregon State University. This topic is about another professor, at Texas A and M University, who is spending part of the appropriation for atmospheric research to check up on mavericks like Chuck Wiese in Seattle.

I received a letter from Associate Professor Dennis M. Driscoll, dated May 13, 1985, saying: "I am conducting research to determine the use that is made of NWS products by television forecasters, and how well these forecasters do in comparison to the zone or city forecasts issued by the NWS. This project, which is funded by a national agency, is nationwide in scope." (Driscoll got my name from Jim Little.)

I said I could supply him with considerable material, but he should tell me about the origin and nature of his research. He replied (condensed):

> I am delighted to learn that you're a holder of the Seal of Approval, of your long experience with telecasting, and that you may be able to help me.
>
> My study, which is financed by a major national agency, has two general aims. First, I'm determining, via a mailed questionnaire (copy enclosed), the extent of modification by telecasters of the zone or city forecasts of the NWS, and asking about the sources of information which prompt such modifications. This questionnaire was sent to 265 seal holders and to one randomly chosen television station in each of the top 210 market areas.
>
> Second, I'm determining how well the weather is forecast by television weather forecasters. This verification

study will compare the forecasts of the NWS and the television weatherperson for about ten large metropolitan areas in the 48 states for a total time of six months.

The questionnaire was as follows (condensed):

1. Do you have access to the forecasts of the NWS?
2. What is the extent of the consideration you give to the NWS forecast?
3. What percent of the time, over the past year, did you transmit a forecast different from that of the NWS?
4. When you have a forecast that was not exactly that of the NWS, which elements were involved? (temp, state of sky, precip statement, wind)
5. What factors are important in influencing you to transmit a forecast different from that of the NWS? [Five factors were listed for selection, but none mentioned decision-making ability.]
6. Do you utilize the forecasts of other agencies (e.g., private forecasting firms)?
7. Indicate length of time you have been a TV weather forecaster.
8. Do you have a Bachelor's degree in meteorology? (Please return by 3-31-85)

On June 5, 1985, I wrote a 2-page reply, pointing out that he had not identified the funding of his research, and I explained some of my views about the domination of public forecasting by the NWS. I enclosed my 36-page Scrapbook to the local meteorological community. This was a mistake, because I never heard from him.

On June 13, 1985 Chuck Wiese called Dennis Driscoll. There was no tape recording, but Chuck called me immediately, and I recorded Chuck's comments. Driscoll freely admitted the research was funded by NASA (National Aeronautics and Space Administration). Why would NASA be interested in these facts?

More than 6 months went by. On January 25, 1986 Larry Schick of KING-TV in Seattle talked with a student from the U of Washington, who was compiling "comparative forecast" scores of KING-TV and NWS for a research project by Professor Driscoll.

On January 30, 1986, Chuck Wiese called the head of the Department of Meteorology at Texas A and M University,

Dr. James R. Scoggins. That conversation was immediately relayed to me, and I recorded Chuck's words. Chuck's immediate recall was approximately as follows, quoting Scoggins:

> I can't tell you.   It's a sensitive situation. They have requested confidentiality. In no way are we going to give this to you until the research is completed, and the results are made available.
>
> If this particular agency revealed itself to the NWS, and told the NWS it was doing this, the NWS would immediately get on the defensive and take the position that this particular agency [the funder] has it in for the NWS, trying to do it in.
>
> You'll be able to know who did the research as soon as it comes out."

The reader is invited to join me in interpreting this limited information as best we can.

On February 3, 1986, I wrote to Dr. Scoggins, sending him copies of my correspondence with Driscoll, Driscoll's admission of funding by NASA, and his (Scoggins) declination to identify the funder. I suggested he reflect on the possibility of embarrassment to his department when the facts are made public. Such letter gave him opportunity to correct any misinformation which I might have.

He did not reply. On May 8, 1986, I wrote again: "It would seem Driscoll's research should now be completed, and that I deserve to learn about it. He solicited my cooperation on May 13, 1985, and he received it."

On May 13, 1986, Dr. Scoggins replied: "Each faculty member who is a Principal Investigator of a research project has full responsibility for that project. Since my knowledge of Dr. Driscoll's research is limited to essentially budgetary concerns, I am in no way qualified to be his spokesman. I suggest you contact Dr. Driscoll directly."

This answer seemed inadequate. A department head can easily forward a public relation problem to a researcher. Furthermore, my inquiry about funding ought to come under budgetary concerns.

It is not that I personally need to know. Serious forecast users and taxpayers need to know. They should not be content with brush-offs.

Five months later, December 29, 1986, I wrote again to the head of the Department of Meteorology at Texas A & M, in which I enclosed the tentative manuscript on this topic. Was there anything he wanted to add?

This time my letter was forwarded to Driscoll, who sent me the galley proofs for his 8½ page research paper to appear in the December 1986 issue of the new AMS publication *Weather and Forecasting*, entitled "A Survey of the Use of National Weather Service Forecasts by Television Weather Forecasters in the United States."

Driscoll explained this was the first half of his research. The second half was very near completion, a "comparison of verification scores between telecaster and the NWS at seven cities in the United States," which he would send as soon as possible.

He wrote: "I trust that in your book . . . you include the results of my survey as well as the survey questions." Excerpts from the abstract follow:

> Questionnaires were mailed to 453 television weatherpersons, and 67% were returned.
>
> Only six percent of all respondents consider the NWS forecast not at all, whereas for almost 60 percent this consideration is moderate to heavy.
>
> The percentages of those who transmit a forecast different from that of the NWS from 0-10, 40-60 and 90-100 percent of the time are 16, 30, and 8 percent, respectively.
>
> The factors which influence the telecaster to issue a different forecast are in order, current weather, local conditions, the telecaster's own analysis of forecast models and his experience, satellite photographs, radar, and colleagues.
>
> Forecasters who are least likely to consider the NWS forecast, and who issue a forecast different from that of the NWS a large percentage of the time, are seal holders with a degree and six to ten years experience who work in one of the top twenty market areas.

Any reader of this book could have deduced that without any research.

In AMS publications, the source of funding for research grants is usually identified, but it was not in this case.

If the National Aeronautics and Space Administration paid for this research I wonder how they will use it. Or maybe someone is misleading us. Also, if scores are being tallied for only seven cities, and maybe that means only seven TV stations, why did Driscoll select Seattle where Chuck Wiese works?

I wrote to Dr. Driscoll on January 26, 1987, again asking for clarification of these matters. I also wrote: "I have looked up your reference by John Walls. (*National Weather Digest*, February 1977, Media forecasts — fact or fiction). How can you suggest such a shallow propaganda piece as a backup for your research?"

I don't see how any comparative verification scheme can be designed on a statistical or arithmetical basis when the formats, and also the objectives, vary so much, between a TV weathercaster and a NWS forecaster. Quality in such composite efforts can only be rated subjectively.

Even the performances of gymnasts and exhibition ice skaters, in carefully defined contests, are rated subjectively by expert judges. I hereby label this NASA research project a waste of funds. I wonder if NASA will raise hell with Texas A & M?

In another vein, with respect to academia, I sympathize with the indignity of having to jump through hoops in order to obtain research grants. If SFUs will support appropriations for legitimate atmospheric research (and I'm sure they will), academia can ignore the federal bureaucrats.

We need Walt Disney to make another movie, this time a comedy about the National Weather Service.

People love to gamble. Weather forecasting is more fun than lotteries or slot machines.

Television program managers will wake up to realize they have a new advertising gold mine.

I am not trying to banish radio and television weathercasters. I am urging them to disseminate genuine forecasts, and to learn how to explain the basis for those forecasts. The NWS is selling them short.

11

Weather Tomorrow Society

This society is yet non-existent, but the 50,000,000 serious forecast users in this nation are hereby invited to become charter members. There will be no requirements for membership except a serious interest in the weather tomorrow. Maybe we can get financial support from established organizations which have much to gain from improved weather forecasts.

About 40,000,000 of these SFUs are voters. Just think of the goodies we can persuade Congress to serve up. We made real progress after World War II, and the nation got back much more than things cost, that is, for a while.

If we get back on track, and with the modern techology which is still exploding, progress in the immediate future can be astounding.

In our new society, which will far eclipse the American Meteorological Society, and throw fear into the ranks of National Weather Service bureaucrats, we will welcome any meteorologists and members of academia who are interested in forecasting.

Any other "experts" who might join will be watched closely for disruptive tactics, because this society will have political goals as its main objective.

Most SFUs are not weather experts, in the sense of academic background, but many of them are experts in some other field. After all, doctors and lawyers go to skilled mechanics when their car needs fixing. We need not stand in awe of insecure professional meteorologists.

If the sign-up lags for the Weather Tomorrow Society, maybe we will change the name to the Weather Gambling Society. Maybe that will better convey its purpose. No doubletalk in this outfit!

One early project ought to be to persuade the University of Washington, the alma mater of Jack Capell, to give him an honorary doctorate degree. He has done more than anyone to bring professional forecasting to television viewers in the Pacific Northwest.

Professor Mass mentioned the need to minimize tensions between federal and private forecasters. Jack Capell has long since earned an HD, doctor of harmony. He also has earned a P by anyone's standards, and that stands for Professional. That adds up to PhD in more ways than one. Jack is not a curmudgeonly crank.

I may sound like a raw political activist. That I am. It all happened within the last 3 years. As already stated, I am a copycat. I learned my new role from studying the National Weather Service. If those bureaucrats can succeed that well by using politics, it is time for serious forecast users to give them some competition.

Strong and free political power. Of the hard-working people, the productive people in the country. This is the American Way. It made our nation the magnet that attracts so many immigrants, legal and otherwise.

A Sober View of the Immediate Job Ahead

In order to break the stranglehold of the National Weather Service on weather forecasting, it must be forced out of its presumed function of public forecasting. Funds must be cut off. It should keep on doing what it is doing now, which is measuring and reporting weather conditions, compiling data, and producing computerized guidance material.

But the public needs an interpretive forecast also, which it mistakenly believes it is getting now, and which it will want after the stranglehold is broken. Where will it get such forecasts?

There are some good independent forecasters available. I have little or no knowledge about them specifically. I can guess their numbers are few. A sudden demand for production probably will overload their systems.

The scene will be crowded with new entrepreneurs, and some of them will be unqualified, and unskilled. The forecast user should not be dismayed, but he or she should be careful.

The evaluation of forecast accuracy is usually called

"verification." Like the evaluation of quality products in other areas of life, forecast verification is not always easy. It may take as much effort as the original forecast.

The so-called "scientific analysis" of NWS forecasts, claimed by Murphy, Douglas, and others, is more biased and self-serving than it is scientific.

The subject is large, and can't be adequately explained in this book, but some abbreviated comments will guide the reader.

Verification of Weather Forecasts

Be skeptical of "objective" systems for verification. They depend on arbitrary rules invented by the verifier, and in the case of the NWS, we get back to the idea of the foxes explaining how the chickens disappeared.

Forecasts must be judged subjectively, the way we select our friends, our mate, buy a car, choose from a menu, etc.

Remember that persistence dominates many weather situations. A blind clerical decision to forecast for tomorrow the same as what occurred today, will produce a "score" of from 60 to 80 percent, depending on the location and time of year. All real forecast skill must exceed that of persistence in order to be counted.

The idea of a hit or a miss is childishly over-simplified. For practical purposes, there is no difference between no rain, and .01 inch, but that is the hit-or-miss line of Murphy and the NWS in scoring rain forecasts.

Any verification scheme must have categories of error, such as a hit, a near miss, a medium miss, and a bad miss (to the extent of evaluation needed).

I have never heard of such a scheme, except the one Howard Graham and I performed on fireweather forecasts in the Pacific Northwest, while working for the U. S. Forest Service. The analysis didn't make the NWS look good. Possibly to avoid embarrassment to the NWS, only 200 copies were printed, of which about 20 were given to me. Later, a NWS spokesman disparaged the analysis.

The idea of scores in percentage, such as 80%, is ambiguous. The rules are seldom or never stated. A score of 100% is presumed to be perfect, a bulls-eye. Dr. Murphy's rule for a rainy day also includes one during which only .01 inch was measured. If the forecaster predicted heavy rain, give him 100% for that day.

If he predicted no rain, which would have been a pretty good forecast, give him zero. So we should laugh at these percentage scores for rain forecasts.

National Weather Service statistics on forecast accuracy are nearly always limited to occurrence of rain, or the temperature. But there are many other weather elements. With regard to life and property, wind is much more important, but we never see any statistics on verification of wind forecasts.

To judge forecasts subjectively (which seems beyond the capability of some statisticians) one must make comparisons. How did the actual forecast compare with a forecast of persistence? How did Chuck's forecast compare with the forecast made by the Seattle WSFO?

The best appraisal of all is to focus on the times when large changes occur, those days only. How did the forecast compare with the large change which was impending?

Local WSFO forecasters merely follow the broadscale generalized prognosis from the central computer in Washington D. C. They function as clerks not as forecasters.

On the majority of days the forecast is for little change, which is good, because persistence is a large built-in factor in day-to-day weather. If the robot can see a change, and it does well with broadscale not-too-fast change, the local clerks will predict a change.

However, if an event occurs quickly, bringing large change, the NWS will miss it every time. Then the Weather Tomorrow Society should say, "Where were you when I needed you?"

The NWS covers up its errors by focussing attention on **watches and warnings** programs, that is, its skill in weather observing. Richard Douglas boasted about their false-alarm ratio. Such scores ought to be 100% all the time, because the warning is made after the event is observed. Such coverup is difficult for the public to understand.

Ambiguity and Obfuscation as a Way of Life

Prediction of the future is an exciting and daring task for any mortal. The psychological armor strapped on daily by weather forecasters is a measure of their nagging fear of meeting John Wayne.

The word nowcasting should always ring a bell. Attention to present weather is **observing**. Communicating the information is **reporting**. The word **nowcasting** is a disguise to offer observing as an easier substitute for **forecasting**. Nowcasting is the transvestite of the weather forecast world.

Consider the word "mostly." I counted the word 25 times in a videotape of a weathercast. The omission of this useless word will not change any meaning. But the timid forecaster who says Mostly Sunny wants to be right on both Sunny and Partly Cloudy, and maybe some Cloudy thrown in.

Consider the spectrums predicted for windstorms. On October 2, 1962, the Portland forecaster predicted 20 to 40 with gusts to 60. Portland suffered gusts to 116. He didn't quite stretch far enough.

On November 13, 1981, Jerry Youngberg predicted 35 to 45 with gusts to 55. The gusts reached 90. Why bother with the 35 to 45? That is to cover your backside if the storm misses you.

Dr. Allan Murphy keeps urging the percentage probability in forecasts for rain. The Portland WSFO only puts them in some of the rain forecasts, and most of the probabilities are for 20 or 30 or 40 percent. We seem we get a lot of 100% occurrences after those 30% probabilities.

Murphy asserts NWS forecasters are ill-prepared to be decision-makers, a surprising admission. And, I will add, they are good at hedging.

The High Cost of Phony Forecasting at WSFOs

There are about 52 Weather Service Forecast Offices, and possibly the personnel roster will vary slightly from one to another. This analysis is based on the roster at the Portland, Oregon, WSFO.

Considering only forecasting personnel and executive positions, there is one Area Manager (Meteorologist in Charge) at grade GM-15, and one assistant at GM-14. There are 5 GS-13 Lead Forecasters, 6 GS-12 Forecasters, and 4 GS-11 trainee forecasters. Within each grade there are 10 salary steps, mostly dependent on length of service, although step increases are awarded for quality performance.

This tabulation will use the mid-point of each grade salary range, although actual salaries on the average probably lie above the mid-point.

The following estimate is for direct annual salaries at the Portland WSFO for forecasting and supervisory personnel, not including contributions to retirement fund by employer, or payroll administrative expenses, or other overhead.

Grade	Step 1	Step 10	Using mid-points each salary range			
11	$27,172	$35,326	4	@	31,249	$124,996
12	32,567	42,341	6	@	37,454	224,724
13	38,727	50,346	5	@	44,537	222,685
14	45,763	59,488	1	@	52,626	52,626
15	53,830	69,976	1	@	61,903	61,903
					Total	$686,934

This estimate does not include salaries for other personnel, such as electronic technicians, one hydrologist, observers, and clerks. It does not include rent, communications, equipment, supplies, or other overhead expenses. The River Forecast Office is independent, and located downtown.

The following information took 3 months and 5 letters and $42.81 to extract from the Western Region of the NWS at Salt Lake City. The request was for figures on the annual operating expenses for the Portland WSFO. The Budget Officer supplied printouts of "Report of Current Year Cost and Obligations by Organizational Element (as of September 30 for fiscal year of 1985)." Jon J. Kelker wrote:

> Costs itemized under "R REIMB" are expenses reimbursed to NWS by other federal agencies and are, therefore, not considered as true operational costs to the WSFO.
>
> Any costs listed in object classes over 42-- are overhead charges of NOAA and the Department of Commerce and are not charged to the WSFO or Western Region.

A request was made for definitions of accounting abbreviations, which brought some definitions but not all, plus the following comment:

9849, 9856, 9857, 9876, 9877 - These object classes represent applied overhead charges for support of NWS operations by NOAA and DOC. They are used for accounting purposes only and do not represent real dollar costs to the
NWS Western Region specifically or NWS in general.

My request for explanation of terminology was renewed
for the third time, "or must I write to my Congressman?,"
which brought some of the remaining answers.

In the table below minor items are not listed, and some
of the abbreviations remain unexplained, but the major items
and the totals are listed. One does not need to be an accoun-
tant to take note of the cost to the taxpayers for the annual
operation of just one Weather Service Forecast Office.

APPROPRIATION R REIMB.		
Reimbursed to NWS by other fed. agencies		$5,351.56
APPROPRIATION 8 OR & F (S & E)		
Personal compensation	$876,933.87	
(plus)		
Overtime	6,629.02	
Night differential	22,254.87	
Sunday differential	20,404.19	
Holiday pay	16,298.00	
Leave Surcharge	165,740.40	
Employee Contrib. Surch.	117,666.48	
Telecommunications	35,962.09	
Depreciation, ADP Equipment	29,762.26	
Deprec. Telecomm. Equip.	15,890.60	
Common Service, all other	69,418.95	
EXAD-OTH-all other	53,260.50	
EXAD-OTH-personal services	102,293.86	
General Support-OTH-all oth	17,043.03	
General Support-OTH-per svc	47,577.87	
General Support-OWN-oth obj	22,393.37	
General Support-OWN-per svc	101,736.64	1,891,818.71
APPROPRIATION 9 MGT. FUND		
Utility Services	13,629.94	32,610.45
Grand total		$1,929,780.72

The sums for object classes "over 42--" (which are the overhead charges of the National Oceanic and Atmospheric Administration, and the Department of Commerce, for the Portland WSFO) totalled $768,179.60, which is 40% of the total expenses of $1,929,780.72 in fiscal year 1985.

The Budget Officer claimed the overhead charges are not charged to the Portland WSFO or the Western Region of the NWS. Then why are they listed in the computer printout for the "Cost and Obligations" for Portland?

Taxpayers must be paying it somewhere. We who are not accountants must assume this $768,180 is related to the cylindrical administrative structure of the NWS.

The items listed for all salaries (not just forecasters) at Portland (including surcharge for pensions) totalled $1,236,138.49. If all forecasters were removed from the roster, and the Area Manager (with lesser duties) reduced to Grade 12, there would be a saving of about $900,000 each year for salaries alone.

This book is addressed primarily to serious forecast users (SFUs). I claim SFUs are denied the best possible public forecasts because the National Weather Service maintains an unfair monopoly on public forecasts. It hinders independent forecasters who might compete with bureaucratic forecasting.

Some of my helpful advisers on this book are not SFUs. They point out that the majority of citizens (the non-SFUs) are not personally interested in the quality of public forecasts. They don't care who makes forecasts, or who makes the most accurate ones.

The Weather Tomorrow Society should convince the non-SFUs that they also have self-interest in this matter. Why should this 70% majority pay a 70% share of the federal cost of maintaining a phony forecasting service, when they seldom, if ever, need any forecasting service?

If non-SFUs are indifferent, and generous enough, to continue present appropriations, then why not redirect such funds to scientific research for local forecasting?

The universal benefits of accurate forecasts will improve efficiency in human affairs, thus increase profits for entrepreneurs, increase employment, and lower prices for consumers. **Property will be protected, and lives will be saved.**

Summary of Purpose of Weather Tomorrow Society

To encourage the production of good forecasts, the Weather Tomorrow Society must establish a system of reward and punishment, as in any other area of human endeavor. The accuracy of forecasts must be critically evaluated. The WTS should incessantly ridicule imposters.

Without any disparagement of weather as a hobby, WTS must avoid a hobby-club atmosphere. It must avoid popular science whiz bang, but encourage a wider understanding of elementary physics, which is about all one needs to understand weather. It should discourage "mystique" shop-talk, which often is used to keep out the outsiders.

In spite of its disciplined search for truth and devotion to purpose, the WST will be a sanguine organization. It will emphasize enthusiasm, vitality, imagination, and creativity.

The 50,000,000 serious forecast users in this nation are hereby invited to become charter members (in the Weather Society Tomorrow). There will be no requirements for membership except a serious interest in the weather tomorrow.

With the modern technology which is still exploding, progress in the immediate future can be astounding.

In our new society, which will far eclipse the American Meteorological Society, and throw fear into the ranks of National Weather Service bureaucrats, we will welcome any meteorologists and members of academia who are interested in forecasting.

This society will have political goals as its main objective.

The National Weather Service must be forced out of its presumed function of public forecasting. Funds must be cut off.

Forecasts must be judged subjectively, the way we select our friends, our mate, buy a car, choose from a menu, etc.

To judge forecasts subjectively, one must make comparisons. The best appraisal of all is to focus on the times when large changes occur, those days only.

Local WSFO forecasters function as clerks, not as forecasters. If an event occurs quickly, bringing large change, the NWS will miss it every time.

The psychological armor strapped on daily by weather forecasters is a measure of their nagging fear of meeting John Wayne.

The universal benefits of accurate forecasts will improve efficiency in human affairs, thus increase profits for entrepreneurs, increase employment, and lower prices for consumers.

The Weather Tomorrow Society should incessantly ridicule imposters.

12

Another Nail in the Coffin

On October 15, 1986, the National Weather Service in Seattle held a Disaster Preparedness Workshop for the King County Office of Emergency Service, the Snohomish County Department of Emergency Management, and the Seattle Department of Emergency Management.

The advance announcement exclaimed — "**Watch Out for Windstorms!** On Friday, October 12, 1962, Columbus Day, the Pacific Northwest was hit by a strong windstorm. This storm claimed 46 lives, blew down 15 billion board feet of timber, damaged 53,000 homes and destroyed thousands of utility poles. Red Cross records show 317 persons hospitalized. Hardest hit was Western Oregon. If a similar windstorm hit Western Washington, how well are we prepared?"

A major winter windstorm **Scenario** was presented at the day-long event at the WSFO. Because "an event like this hitting the Puget Sound would **impact the lives of over 2,000,000 people**" the NWS put on this propaganda circus for local emergency organizations and the media.

This book makes assertions about the incompetence of NWS local forecasting, and presents evidence for substantiation. The most convincing evidence is that with which the NWS reveals itself. Serious forecast users must bury the phony public forecasting offered by the NWS. This circus pounded another nail in the lid of the coffin.

A scenario is an outline of the plot of a dramatic story, a hypothesized chain of events. Apparently, the NWS assembled material from files describing the windstorms of 1962 and 1981. Apparently, my criticism of the Charles D. Mitchell Award for the storm of November 13, 1981, was ignored because that poor performance was blended into the scenario.

Such insensitivity reinforces the claim that the NWS doesn't understand that prediction involves the future, and that the NWS thinks and acts like an observer and reporter. The scenario will be reproduced in its entirety, with slight non-critical explanations added.

Although weather events are easier to understand than most people believe, there are some inherent complications. The events involve multiple elements such as wind, pressure, temperature, etc., plus varied geography, plus the sequence of events in time. The serious forecast user must keep the story orderly in his mind.

The scenario covers a period of 51 hours, beginning at 9 p.m. Wednesday, October 15, 1986, the day of the workshop, and ending at midnight Friday night, October 17. But the NWS got distracted by events listed for Friday afternoon and changed the date to the 18th at 5 p.m. You see, even the author had difficulty in maintaining his continuity.

Instead of day of week, day of month, and clock time, the hours here are listed merely from 0 to 51. My comments inserted within brackets.

Hour   0 TV weathercasters briefed by NWS. Concern of small surface wave, a small depression in the Eastern Pacific well off the California Coast beyond 140 west longitude. Upper level pattern favors storm development for the Pacific Northwest with a strong jet stream dipping south of 40 north latitude. Meteorologists call this cyclogenesis.

Hour   7 Forecasters with satellite loops see that cyclogenesis is taking place.

Hour 12 Coast Guard sent on search and rescue mission off Oregon Coast.

Hour 14 Computer models move a surface low toward the Oregon Coast in 48 hours. If timing is off, and if this is the same surface low, Western Washington will be spared from the major effects of the storm with most of the energy directed into Oregon and California . . . but the models have had a south bias lately.

Hour 16 The surface low moves into zone "A" [a little larger than the areas of Washington and Oregon, centered 730 miles WNW of San Francisco]. From past history this is an area in which storms are centered 24 hours prior to a high wind event in Puget Sound.

Hour 17 Coast Guard rescues crew from small freighter sinking off Oregon Coast in rough seas.

Hour 19 Meteorological Buoy 06 reports an 8 millibar fall in past 3 hours. [the buoy is 690 miles west of Crescent City, California]

Hour 21 Surface low near 40/135 may be undergoing explosive deepening [intensification] at the rate of over 1 mb per hour.

Hour 22 High Wind Watch issued for all of Western Washington. NAWAS alert statement sent over NAWAS to all emergency services agencies and sheriff offices to all the counties West of Cascades.

Hour 23 Special Weather Statement issued. Radio and TV stations carry the bulletin during the late evening newscasts. This is a potentially dangerous storm, and the public is advised to be prepared for possible power outages and some wind damage if this storm keeps moving at its present rate and should be into the Puget Sound area by Friday afternoon [1 p.m. would be Hour 40, when Scenario begins high winds there]

Hour 24 New forecast charts move the surface low into Vancouver Island in 24 hours.

Hour 25 Surface pressure rises along the Northern California coast, while the low moves northward off the coast toward Oregon. Pressure falls over Pacific NW show greatest falls on the Central Oregon coast. Could this be where the low is moving? Strong southeast surface winds are noted from ship reports off the South Oregon coast.

Hour 26 Surface low continues to undergo explosive deepening. Central pressure has fallen 11 mb in past three hours. Low is now located at 41.5N and 129W, just off Crescent City, California [255 miles east of storm center].

Hour 28 Center of storm continues to deepen, now near 930 mb [wind storms of 1962 and 1981 had central pressures of about 960 mb], and now seems to be moving toward Southern Washington coast. BAROGRAPHS are falling all over Western Washington. Further analysis indicates that the center may be heading toward Vancouver Island.

Hour 31 Gale Warnings issued for all inland waters of Western Washington. SOUTH WINDS OF 40 KTS can be expected by early afternoon [Hour 40]. Sea-Tac Airport and Boeing Field alerted for high winds.

Hour 32 Special Weather Statement from NWS sent over NAWAS . . . NOAA Weather radio and over Weather Wire to news media. This storm is potentially very dangerous and winds of 40 mph with gusts of 60 mph are possible in the Puget Sound lowlands later today.

Hour 33 Radiosonde winds from Salem, Oregon show that there were winds aloft at 4000 feet of 47 knots.

Hour 34 Barograph traces in Central Oregon have begun to show pressure rises, while falls are still occurring at the Seattle Forecast Office. Two men make plans to take 48 foot sailboat out in strong winds this afternoon on Shilshole Bay.

Hour 35 High winds have started in the Willamette Valley of Central Oregon with southerly gusts of 60-70 mph.

Hour 37 Radiosonde lost on ascent from Portage Bay Meteorological Station. Wind shear forecasts issued for all airport terminal forecasts in Puget Sound. High winds of over 60 mph were reported from Coast Guard Stations on South Washington coast.

Hour 38 Special Weather Statement and High Wind Warning issued by NWS.

Hour 40 [1 p.m. in Scenario] High winds begin in Puget Sound. Light southeast winds at Sea-Tac suddenly shift to the south with speed of 35 knots and gusts to 65 mph. Wind reports from other airports are a little stronger, with Renton reporting gusts to 75 knots. Radio HAM reports 65 mph gusts in Tacoma.

Evergreen Bridge reports gusts to 100 mph. EVERGREEN, LAKE WASHINGTON, TACOMA NARROWS, AND HOOD CANAL BRIDGES ALL CLOSED. TV tower on Gold Mountain reported blown down.

Power outage widespread. Emergency power kicks on at the NWS Forecast Office. 75% of King County reported to be without power. 65% of Pierce County without power. Ship has dragged its anchor on Elliott Bay.

Several people reported injured in downtown Renton. Many homes in King and Snohomish County have roof damage. South-facing hillsides reported severely hit by downed trees in many foothill areas. Large blowdown of trees in parts of Snohomish County. Many roads throughout King County have downed trees and power lines.

Four people lost on Lake Sammamish — overturned boat. All ferries in Puget Sound suspend operations. Gusty winds and high tide. Damage reported high at Shilshole. Tidal flooding reported in Kitsap County near Poulsbo.

Space Needle elevators stop — several people stuck midway up. Windows blew out in high rise buildings in downtown Seattle and Bellevue. 100 people injured in Seattle from glass and debris.

Large crane on new building in Bellevue collapses. 90% of power reported out in Everett and Snohomish County and the same for Skagit County. Two sailors in large sailboat reported lost overboard near Gedney Island near Everett.

Hour 44 Winds continue to blow from the south at 35 kts in Seattle area . . . shifting to the southwest at times.

Hour 46 Wind gusts of 50 kts still occurring in some areas.

Hour 48 Surface low has now moved into Western B.C. Winds have decreased and now from the southwest at 25 mph.

Hour 49 High Wind Warning cancelled. Extensive damage reported over Puget Sound lowlands. Some power has been restored, but over 65% of Puget Sound remains without power. Trees are down over much of the county roads. Structural damage reported on many buildings in Marysville. Dairymen concerned on milking--no power.

Hour 51 [midnight] Winds finally are down to southwest 12 to 18 mph. Power crews work through the night restoring power to many residential areas.

Panelists are requested to address the following two questions: 1. How did you get the information? 2. What type of action is required by your agency?

Although the scenario was fictional it presumably was based on historical facts, and the attitudes and policies of forecasters at the WSFO. What was revealed? Keep in mind the invited audience was composed of people who were being urged to accept, act upon, and admire the "products" of the NWS.

The NWS WSFO is not a general news-gathering agency. If its primary mission is the saving of life and property, then emphasis must be placed on forecasting, rather than subsequent reporting of weather events and the enhancement of such reports under a label of Special Weather Statements.

The main Special Weather Statement the media needs, and the public needs, is an accurate forecast as early as possible. After the storm, let the news agencies summarize the loss of life and property.

Consider those parts of the scenario which relate to the presumed predictions of the NWS. Also remember that a "Watch" is not a forecast, it is just a "maybe" that something "might" happen, or maybe it might not. A Watch is part of the psychological armor of a timid forecaster.

These comments are directed at specific Hours in scenario:

7. A satellite loop is a series of cloud pictures displayed in sequence on a videotape to simulate a time-lapse movie — interesting but non-essential. The sequence of surface weather maps at 6 hour intervals, and upper air charts at 12 hour intervals is the framework of weather analysis. Modern forecasters in the NWS seem to have lost the ability to study such charts.

17. If ship is sinking, it is not related to a storm 700 miles from the nearest coast of Oregon.

19. Rate of pressure fall at any location is indeed a fundamental indicator. Pressure is now measured in millibars, with 1014 the average everywhere at sea level. A standard period of 3 hours is used to determine the "tendency," shown here as 8 mb, an impressive rate of fall. A fall of 2 mb in 3 hours will attract attention. During the 1962 windstorm falls of 10 were measured, which caused this author to tremble with fear.

The primary reason for pressure fall at a given location is the approach of a storm. If the storm itself is intensifying, meaning the central pressure is lowering, such factor is added.

21. Intensification is unmistakable, not maybe.

22. The emergency services are now entitled to more advice than reassurance the NWS is watching the developing storm. This is time for a specific forecast, even if it might be amended later.

   The upper air charts, not mentioned here, clearly indicate the storm will steer far enough north to bring a major windstorm to Puget Sound. As an approaching storm intensifies, the upper winds ahead of the storm "back", meaning the direction shifts in a counter-clockwise direction, as from WSW to SW to SSW.

23. Advice about possible power outages and some wind damage is not the job of a weather forecaster. He should predict the peak gusts and time of arrival.

24. Reading prognostic charts aloud is a cop-out. The local forecaster is responsible for the forecast. That is what he is paid for.

25. Obviously the storm has been moving toward the place with the greatest fall in pressure. Obviously the path will curve gradually to the left.

26. Gary Cooper is about to confront the forecaster at High Noon, but apparently the forecaster is not yet aware of this.

28. A storm of this intensity would well exceed any known windstorm in the Pacific Northwest. A real forecaster would be scared to death.

31. The pantywaist forecaster issues Gale warnings, which cover the range of 39-54 mph. For suburbanites, this is "garbage can" weather. Move it into the garage. Maybe the forecaster is waiting for Washington D.C. to tell him what to do next.

32. Nothing Special about this statement, with words "potentially" and "possible." Gusts should be predicted to at least 120 mph, which will produce about 8 to 10 times the damage of gusts to 60.

   Gusts are caused by small masses of air decending to the surface because of turbulence. They immediately are slowed by friction. They never are accelerated. The wind speed in the free air, above the friction effects, is fairly uniform, but highly variable in the friction layer.

If any spot measures a gust of 100, the winds above are at least 100, and hence many other spots probably also get gusts of at least 100. The so-called "observed" peak gusts during a storm are obviously incomplete measurements.

35. Under the scenario described, the winds in Willamette valley would be blowing 140 in gusts, not 60-70. The Willamette Valley (with accent on "lamb") with a population of over one million, is near the western edge of Oregon.

37. Wind shear is a recent buzzword invented by those who have just discovered thunderstorm downdrafts, which have been understood by fireweather forecasters for about 50 years. This phenomenon, simple but devastating, has been involved in several recent aircraft accidents.

Thunderstorm downdrafts are not a focus of concern in the Pacific Northwest windstorms of the Columbus Day type. The scenario is indulging in popular science whizbang.

38. This book asserts that NWS warnings are made after the event is observed. That does not necessarily mean at the WSFO, but nearby and moving toward the WSFO. With its communication network, financed by taxpayers, the NWS can see beyond the visual horizon.

One hour earlier high winds of **over** 60 mph were reported on the South Washington coast. How much over 60 mph?

At this workshop the NWS was talking to key emergency and media people. The scenario must be taken at face value. The first high wind warning was issued when the winds were coming over the hill.

The lengthy scenario under Hour 40 suggests the attitude of a reporter. A reporter is not faced with uncertainty about the future. He is a spectator. He doesn't have to face John Wayne or Gary Cooper. After the shootout he will tell you about it. This is why old forecasters have scar tissue. Often they are wounded while learning to shoot fast and accurately.

Eventually they become tough, as in curmudgeonly old crank.

Robin Cody's question has been answered.

WHY CAN'T THEY GET IT RIGHT?

Where Do We Go From Here?

We won't go anywhere as long as the National Weather Service maintains its monopoly on public weather forecasting. It is not likely to yield that monopoly until its appropriation for forecasting is eliminated.

Serious forecast users should expect a vigorous propaganda campaign from the bureaucrats in the NWS.

Amid the coming shakeup, a reader of this book can more skillfully translate the impending barrage over the airwaves.

March 5, 1987